UNVEILING GALAXIES

The Role of Images in Astronomical Discovery

JEAN-RENÉ ROY

CAMBRIDGE
UNIVERSITY PRESS

CAMBRIDGE
UNIVERSITY PRESS

University Printing House, Cambridge CB2 8BS, United Kingdom

One Liberty Plaza, 20th Floor, New York, NY 10006, USA

477 Williamstown Road, Port Melbourne, VIC 3207, Australia

4843/24, 2nd Floor, Ansari Road, Daryaganj, Delhi - 110002, India

79 Anson Road, #06-04/06, Singapore 079906

Cambridge University Press is part of the University of Cambridge.

It furthers the University's mission by disseminating knowledge in the pursuit of
education, learning, and research at the highest international levels of excellence.

www.cambridge.org
Information on this title: www.cambridge.org/9781108417013
DOI: 10.1017/9781108261104

© Jean-René Roy 2018

First published 2018

Printed in the United Kingdom by TJ International Ltd. Padstow Cornwall

A catalogue record for this publication is available from the British Library.

Library of Congress Cataloging-in-Publication Data
Names: Roy, Jean-René, author.
Title: Unveiling galaxies : the role of images in astronomical discovery / Jean-René Roy
Description: New York, NY : Cambridge University Press, [2018] |
Includes bibliographical references and index.
Identifiers: LCCN 2017023077 | ISBN 9781108417013 (alk. paper)
Subjects: LCSH: Astronomical photography. | Galaxies. | Astronomy.
Classification: LCC QB121.R697   2018 | DDC 523.1/12 – dc23
LC record available at https://lccn.loc.gov/2017023077

ISBN 978-1-108-41701-3 Hardback

UNVEILING GALAXIES

The Role of Images in Astronomical Discovery

Galaxies are known as the building blocks of the universe, but arriving at this understanding has been a thousand-year odyssey. This journey is told through the lens of the evolving use of images as investigative tools. Initial chapters explore how early insights developed in line with new methods of scientific imaging, particularly photography. The volume then explores the impact of optical, radio, and X-ray imaging techniques. The final part of the story discusses the importance of atlases of galaxies; how astronomers organised images in ways that educated, promoted ideas, and pushed for new knowledge. Images that created confusion as well as advanced knowledge are included to demonstrate the challenges faced by astronomers and the long road to understanding galaxies. By examining developments in imaging, this text places the study of galaxies in its broader historical context, contributing to both astronomy and the history of science.

JEAN-RENÉ ROY is a retired astronomer who was a professor at Laval University, Québec, Canada from 1977 to 2000. Since then he has served as Deputy Director and Head of Science at the Gemini Observatory in Hawai'i and Chile and worked at the Large Facilities Office of the National Science Foundation and finally at the Space Telescope Science Institute. He has done research on the Sun, the interstellar medium, and the evolution of gas-rich galaxies. A new edition of his previous book, *A Question and Answer Guide to Astronomy*, was published in 2017.

"As one of the world's leading astronomers, Jean-René Roy provides us with an insightful and readable account of the use of images to distinguish between deep-sky objects, such as nebulae and galaxies. What makes this an exceptional work is the level to which Roy, as a practitioner, engages with historians of science in developing his rich account. This engagement leads to a unique book, one that will be indispensable to understanding the significant role played by images in the history of twentieth-century science."

Omar Nasim, *Universität Regensburg*

"*Unveiling Galaxies* examines the role of 'transformational images' in the history of astronomy. Images are a tool of discovery, and this book brings attention to the groundbreaking images behind some of the greatest discoveries in astronomy. The book also highlights the role of galaxy atlases in astronomy as well as the lives of the people who made these images and how their work impacted the progress of astronomy. I found that telling the story of the discovery of galaxies by focussing on iconic images is an excellent approach to the subject. *Unveiling Galaxies* is informative, well written, and well researched, and provides a superb read of the process of discovery in science."

Ronald J. Buta, *University of Alabama*

"In this beautifully clear, reflective, and almost non-mathematical book, Jean-René Roy explains how we came to understand that galaxies are the building blocks of the universe. Roy is an accomplished galaxy researcher who takes us on this long and fascinating journey with its many challenges, from the perspective of developments in scientific imaging of galaxies. The story is based on images, starting with sketches of galaxies made from visual observations in the 18th and 19th centuries. The book describes how the gradual improvement in the quality of the images led to the understanding that galaxies are very distant objects, lying far outside the Milky Way."

Kenneth Freeman, *The Australian National University*

"Roy's unique contribution goes beyond tracing the development of making images of galaxies to examine their compilation into atlases. Roy's underlying motivation for this work is personal; his own exploration of a gift of an atlas of galaxies sparked his interest in science and astronomy. With the descriptions of the characters who contributed to the progress of understanding galaxies, the author reminds us that science is a human activity. This book touches on the highlights of how images proceeded from eye and hand to photographic and lately electronic record."

Nancy Levenson, *Space Telescope Science Institute*

We may, therefore, well hope that many excellent and useful matters are yet treasured up in the bosom of nature. Francis Bacon

To the memories of Allan Rex Sandage, Gérard de Vaucouleurs and Halton Christian Arp

Contents

Preface

Galaxies are the building blocks of the universe. They have been and continue to be extraordinary objects for probing and understanding the universe, its origin and evolution. This book is about galaxies, and about how images led to their discovery and contributed to the understanding of their nature.

Held together by gravitation, galaxies are gigantic systems of stars and clouds of gas and dust. They populate the universe in the billions. Extrapolating counts from the deepest current observations by telescopes in space and on the ground, one estimates that there are more than 200 billion galaxies in the observable universe. We belong to one of them, the Milky Way, a typical large spiral galaxy: disk-shaped like a slightly inflated pizza, the Milky Way hosts about 200 billion stars and measures more than 100,000 light-years across. Our Sun is one of its numerous stars and is located at about 27,000 light-years from the center of the giant whirlpool of stars, interstellar gas and dust. We have only become aware of our cosmic geography in the last century.

About a hundred years ago, in the late 1910s and through the 1920s – a remarkable decade – astronomers proved beyond doubt that our Milky Way was one of numerous giant stellar systems, and that a multitude of similar "island-universes" were scattered at colossal distances from each other. "Nebulae," noticed by sharp eye telescope viewing all over the sky during previous centuries, were found to be extremely numerous, at least millions in number, and complex in their appearances, from mottled disk-shaped pinwheels to soft spheroidals. The diversity of silhouettes and forms turned out to be a key to understand the formation and evolution of galaxies. Probed by inquisitive astronomers with the modern photographic equipment of the twentieth century, "nebulae" became galaxies and literally opened the door to grasping the whole universe. Some say, we then found the universe.[1] Getting there was a fascinating story.

From the first viewing of a "nebula" in the tenth century to the time these foamy patches in the sky were explained and understood, it took more than a thousand years. Why did it take so long? Discovering galaxies and determining their nature was indeed a long path full of obstacles, confusion, debates, conflicts and finally convergence. In this book I will re-tell this arduous road and exhilarating venture. To do so, I will employ a specific perspective:

---

[1] M. Bartusiak, *The Day We Found the Universe*, New York: Vintage Books, 2010.

the role of viewing, drawing and photographing in figuring out the nature of nebulae, and I will show how this long and time-honored imaging process has helped to bring out the world of galaxies.

Why Images?

An image allows us to see things, to record and describe them, to prove that they exist and to reflect on them. Images can bring irrefutable pieces of evidence for a concept or proof of the reality of a "thing." One can count objects on an image, measure their shapes, identify features and establish if they contain structures, then describe these. When images are digitized, mathematical operations can be executed and quantitative information can be extracted from them. With a large number of images at hand, one can compare and recognize variations between objects. In some cases, such as it was with "nebulae" over the centuries, we might not have a clue about what the "thing" is. Once it is understood, common trends can be traced, "peculiar cases" can be identified and orderly behavior inferred.

Scientists are illustrators. They learn by juxtaposing objects and images. Images then provide an empirical basis for classification, often a critical pre-discovery step, to understand the nature of objects, "nebulae" and galaxies, in our case.[2] Atlases are the juxtaposition of images. Scientific atlases with their rich compendia of images become the pictorial beacons helping us to navigate the natural world. I will explore why and how the great atlases of galaxies of the past decades were put together. I will explain how these image galleries helped to broaden our knowledge of the world of galaxies, how they influenced research programs and drove the design and construction of new telescopes and cameras.

We read text; we read images. Scientific images are not only powerful conveyors of information but also a tool to share complex knowledge.[3] Images also carry esthetic value. They help to create enthusiasm and understanding. Images of astronomical objects can be truly beautiful, esthetically abstract and amazingly representational.[4] The popularity of the Hubble Space Telescope is in great part due to an outstanding and successful educational effort to explain and share particularly significant astronomical images with an audience broader than the astronomer specialist using the facility.[5,6]

My approach and perspective in writing this book are those of a practitioner: it is the viewpoint of someone who has observed and studied galaxies, taught and trained students into learning what they are, helped in building instruments to image galaxies and, finally, run a large observatory where major programs were conducted to explore these prodigious sidereal objects. My career was ignited by a spark. As a teenager, I was

---

[2] S. J. Dick, *Discovery and Classification in Astronomy, Controversy and Consensus*, Cambridge: Cambridge University Press, 2013, pp. 233–276.

[3] M. Lynch and S. Y. Edgerton Jr., Aesthetic and digital image processing: representational craft in contemporary astronomy, *The Sociological Review*, 1987, Vol. 35, pp. 184–220.

[4] M. Benson, *Cosmigraphics, Picturing the Universe Through Time*, New York: Abrams, 2014.

[5] E. Snider, The Eye of Hubble, Framing Astronomical Images, *Frame: A Journal of Visual and Material Culture*, Issue One, Spring 2011, pp. 3–21.

[6] For example, see Z. Levay, *Hubble Space Telescope: Re-imagining the Universe*, TEDxKC, 2015, www.youtube.com/watch?v=JDJsiEI_0gE

dazzled and inspired by *The Hubble Atlas of Galaxies* by Allan Rex Sandage, a giant of twentieth-century astronomy.

To ground my story on a broad foothold, I also borrow the works of scholars who have disentangled the scientific discovery process: historians, sociologists and epistemologists. Their perspicuous investigations have explored the minds of scientists and dissected their products. These scholars have helped us understand why it took 1,000 years from the first viewing of nebulae to the final unveiling of the world of galaxies in the initial part of the twentieth century. Thus, I encapsulate the works of many. Ambitiously, I am striving to build a bridge between the genre of science writing typical of scientists and the history of science literature.

On Nebulae

"Nebulae" are far away, and most are astonishingly distant. The concept of "nebulae" has been at times very confusing, even chaotic, as successive attempts were made to distinguish the categories of these elusive cosmic objects. It took a long time to figure out the diversity of "nebulae" and to understand them. For a long time, astronomers were unable to determine their distances, which many thought implausible. Moreover, the puzzling objects could not be related to anything familiar. The nature of "nebulae" was considered to be out of the ordinary: were they made of a mysterious cosmic fluid, a bunch of unresolved stars or just illusions in the mind of imaginative observers?

To assist you in navigating through this long and foggy history, I give here the gist of the different classes of "nebulae" as we now know them. This will help you to stay the course through the maze of the long-lasting unraveling of their nature.

There are two main classes of "nebulae": (i) diffuse nebulae (clouds of gas and dust) and (ii) extragalactic "nebulae" or galaxies (huge assemblies of stars). Diffuse nebulae are members of our Milky Way; they also exist in other galaxies as components of the interstellar medium – the space between stars. Diffuse nebulae can be divided further into two broad categories, (a) emission nebulae where the atoms of the cosmic gas are stripped of their electrons and made fluorescent by the ultraviolet light of hot massive stars, and (b) reflection nebulae whose dust reflects the light of stars in their vicinity.

By far the largest category of "nebulae" are the non-galactic or extragalactic "nebulae," now called galaxies. Much larger physical entities than emission and reflection nebulae, they are totally different from diffuse nebulae. Galaxies are made of billions of stars and contain huge quantities of interstellar matter often seen as diffuse nebulae and dust clouds. They form two main categories, ellipticals and spirals.

A major epistemological stumbling block was the following: for centuries, most researchers tried to bring all nebulae under one umbrella, making them a single class of physical objects. This is not an uncommon approach in the development of science. The long quest is not without parallel to Plato's allegory of the cave, where people try to understand the world by watching shadows on the walls from the things passing in front of the fire behind them. For centuries, astronomers, like the prisoners of Plato's cave, puzzled over "nebulae." The challenge and key to a successful epistemic exit were to distinguish the different categories of "nebulae," and, as a critical step, to establish their distances. Key breakthroughs came with reliable distance determinations and from spectroscopy. The latter technique revealed the physical nature of sidereal matter in its various states, providing the tool to distinguish stars from true nebular material.

Throughout the initial chapters of the book, I will use the word "nebula" in quotes since historically the object discussed could be either a cloud of gas and dust, an unresolved cluster of stars or a distant galaxy, the observers not knowing or being unable to make the distinction. When unquoted, nebula refers to diffuse or reflection nebulae. More confusing for the unfamiliar reader, galaxies were initially called "non-galactic nebulae," "extragalactic nebulae," or "anagalactic nebulae." After the death of Edwin Hubble in 1953, they became simply galaxies. Just watch for the shifting of names, especially when I cite original material.

Plan of the Book

The book is divided in three parts. In the introduction, I deal with the challenges of images and their role in scientific discovery. I discuss the issue of images not being self-evident. Part I deals with the specificity of astronomical imaging and its challenges at finding and revealing galaxies: the long path from the visual discovery of fuzzy celestial clouds to the photography of multitudes of spirals, a long quest that lighted the path to our finding of the universe. I show how images provided the exacting and essential steps for unveiling the world of galaxies: first from written descriptions of what was seen through the telescope (Chapter 1), then sketched in the drawings of nineteenth-century visual observers (Chapter 2), later photographed by the pioneers of the end of the nineteenth century and early twentieth century (Chapter 3), then abstracted as images for the mind (Chapter 4). Chapter 5 acts as a gateway. It is a transition chapter: I recap the whole story in a more traditional way, bringing together the names of the key actors, their places and dates, as well as the ideas that contributed to the unveiling of the world of galaxies. I describe how galaxies became stepping stones for measuring the size and the age of the universe, and not least can be used to determine the exact position of the Sun and solar system in cosmic space and time. I chronicle the crucial decade of 1915–1925, where reliable distances to galaxies were established. I refer extensively to the works of astronomers, both professionals and amateurs, as the latter often contributed in most innovative ways. For example, the early recognition that imaging techniques (photography and spectroscopy) could be valuable to study nebulae and galaxies came from amateur astronomers. It took decades for the professionals, who initially mistrusted photography, to be convinced.[7]

---

[7] See A. Hirshfeld, *Starlight Detectives: How Astronomers, Inventors and Eccentrics Discovered the Modern Universe*, New York: Bellevue Literary Press, 2014.

In Part II, I summarize the current knowledge about galaxies, emphasizing the role of several imaging techniques that helped to unravel the complexities and extraordinary properties of galaxies: galaxies as viewed in the optical/infrared domain (Chapter 6), and at radio and X-ray wavelengths (Chapter 7). A particularly gripping story is how dark matter was discovered and inferred by what I call "imaging the invisible" (Chapter 8).

Part III of the book is my most original contribution. It is about atlases of galaxies. The scientific atlas is a standard tool to share and disseminate knowledge using carefully selected sets of images. "Scientific atlas images are images at work, and they have been at work for centuries in all the sciences of the eye, from anatomy to physics, from meteorology to embryology."[8] Astronomy is no exception. Atlases of galaxies have been trailblazers in the development and sharing of new knowledge about these great assemblies of stars. Classification of objects is the foundation of any scientific atlas. Chapter 9 tells the fascinating story of the building up of a classification system of galaxies and what role images played in the controversial process, and how morphology became a fundamental criterion to classify galaxies. In Chapter 10, I discuss and review all the major galaxy atlases that are published and used by astronomers and their students. In Chapter 11, using specific examples, I illustrate the impact of these atlases on the way research programs were proposed and conducted, and their role in the design and building of new astronomical cameras and telescopes. I conclude the book with some personal reflections on how images are helping us to understand the universe better, and what great tools they are for sharing that knowledge more broadly. Finally, I reflect on the changing role and future of galaxy atlases in the digital age.

From the very beginning, I wish to highlight the cumulative approach I will take you through as we move along. The main thesis (why it took so long) and theme (images as discovery tools) will gradually bulk up over the course of the book. Concepts, ideas, theories, observations, objects and historical actors will occur in different ways as we progress from one chapter to another. There will be repetitions. As we come back to these notions over and over again, we will build a fuller picture. My goal is that by the end of the book, you will appreciate these things in a more rounded fashion and be able to embrace a deeper perspective than provided by the standard astronomy textbook.

The research for this book is based on a mix of primary and secondary sources. Primary sources include research works published in observatory reports, in journals of professional societies, conference proceedings and galaxy atlases. Secondary sources are other scientific atlases, books and articles by researchers in history, sociology and epistemology of astronomy and natural sciences. The *Biographical Encyclopedia of Astronomers* was a rich source of information and provided many hints for further search. The Smithsonian Astrophysical Observatory/NASA Astrophysics Data System Bibliographic Services has been an inestimable resource. Several Wikipedia articles provided useful content and were indicators for other material.

---

[8] L. Daston and P. Galison, *Objectivity*, New York: Zone Books, 2007, p. 19.

This work was inspired and completed with the help of many individuals with whom I interacted during my career. I owe enormously to colleagues who read the evolving versions of the manuscript and provided most helpful comments, criticisms and suggestions.

I am deeply indebted to Omar W. Nasim of Regensberg University, who generously shared his extraordinary perspective, unique knowledge and insights on the role of image making in astronomy of the nineteenth and twentieth centuries. Pierre Martin (University of Hawai'i/Hilo), Ronald Buta (University of Alabama), and Anton Koekemoer (Space Telescope Science Institute) read initial versions of the manuscript and provided extremely helpful feedback. Pekka Terrikorpi of the University of Turku, Finland, provided clarification on the Hubble vs. Lundmark altercation regarding priority on the design of a proper morphological classification of galaxies. David L. Block of the University of Witwatersrand and Kenneth Freeman of the Australian National University shared key information on the early work of John Reynolds on galaxies, and on his forgotten contribution to what later became the "Hubble sequence."

Zoltan Levy (STScI) instructed me on the subtle art and science of making "portraits" from the images obtained with the Hubble Space Telescope. Lars Lindberg Christensen of European Southern Observatory provided provocative insights on the future of imaging in the evolving world of interactive archives and the challenge of creating "ethically correct colour imagery" with raw data from telescopes.

In my exploration of the impact of atlases of galaxies, I interacted with several people by e-mail and telephone. I am most grateful to Alar Toomre (MIT) and François Schweizer (Carnegie Observatories) for their wonderful recollections on the development of the concept of interacting galaxies and of the impact of Halton Arp's *Atlas of Peculiar Galaxies* on their own thinking. I also extend my thanks to Wendy Freedman of the University of Chicago, Kenneth Freeman of the Australian National University, Marshall McCall of York University, Preethi Nair of the University of Alabama, Robert J. Hanisch of the National Institute of Standards and Technology, Eduardo Hardy of Associated Universities, Inc. and Harold G. Corwin, Jr. for sharing on how galaxy atlases did influence their work.

Communicating with librarians around the world has been a most rewarding experience. I thank Jill Langstrom, Head Librarian, and her team at the Space Telescope Science Institute, and Xiaoyu Zhang, the Gemini Observatory Librarian. I am particularly grateful to Paul Espinoza, Curator of the magnificent George Peabody Library, Johns Hopkins University, for hosting me several times and providing me with the publications of the Earl of Rosse. Professor Earle Havens, the William Kurrelmeyer, Curator of Rare Books and Manuscripts of the Sheridan Libraries of the Johns Hopkins University, shared his passion for ancient scientific books and introduced me to the glory of JHU collections of remarkable books. Cynthia Hunt, Social Media Strategist at Carnegie Observatories, and Chair of the History Committee for the Carnegie Observatories, provided important images, and permissions to reproduce them. I owe much to Daina Bouquin (Acting Head Librarian) and Maria McEachern at the John G. Wolbach Library of the Harvard-Smithsonian Center for Astrophysics, Harvard University. They gave me access to several key nineteenth- and early twentieth-century works and references, while shoring up generous and most professional assistance in identifying images for digitization. Loma Karklins of Caltech, Lianne Smith of King's College London and Janet Laidla of University of Tartu Museum came up with notable illustrations. Several librarians helped me to identify sources of images and to clarify ownership of material. Some individuals provided me with hints for other images or different versions, and help in identifying the exact sources. I am most grateful to Mark Bailey of Armagh Observatory and Planetarium, Jaan Pelt of Tartu Observatory, and Harold G. Corwin Jr.

I am in unbounded debt to Matt Mountain for inviting me to spend a year working in the Science Mission Office of the Space Telescope Science Institute, Baltimore, MD. This opportunity and multiple exchanges with STScI colleagues led me to reflect about scientific images and their function in astronomical discovery. My stay at the institute launched me into a small project on the role and impact of atlases of galaxies and that ultimately evolved into this book.

I acknowledge the unrelenting and most professional support of Vince Higgs and Lucy Edwards of Cambridge University Press throughout all phases finalizing the manuscript. Zoë Lewin, my copyeditor, deserves special recognition for her superb work.

Hélène Allard, my companion of decades, remained by my side all the time and patiently reviewed and criticized the earliest versions of the manuscript.

Introduction

As the dome shutters begin to close, they [coyotes] emit a high-pitched squeal that every coyote within a three-mile radius answers with a howl. Their primeval lamentations play a fitting coda to a night of solitude with the stars, the dome, and the slowly turning telescope.

Allan Sandage[1]

To learn to observe and to depict in a science is to acquire at once an ethos and a way of seeing.

Lorraine Daston and Peter Galison[2]

In one sense, that theory of the spiral nebulae to which many lines of recently obtained evidence are pointing, cannot be said to be a modern theory. There are few modern concepts which have not been explicitly or implicitly put forward as hypotheses or suggestions long before they were actually substantiated by evidence.

Heber D. Curtis[3]

How Are Images Discovery Engines?

Images of galaxies are at the core of this book (Fig. 0.1, see Plate 6.3). I set the course by showing how images have been used for discovery and to further research, with a particular emphasis on their link to knowledge and trailblazing roles in the unearthing of natural processes.

Troublesome Images

Images that we so naturally use as conveyors of information were not always accepted so easily. Walking nowadays along the streets and boulevards of bustling Istanbul in modern

---

[1] A. R. Sandage, *Centennial History of the Carnegie Institution of Washington, Volume I: The Mount Wilson Observatory*, Cambridge: Cambridge University Press, 2004, p. 195.
[2] L. Daston and P. Galison, *Objectivity*, New York: Zone Books, 2007, p. 367.
[3] H. D. Curtis, Modern Theories of the Spiral Nebulae, *Journal of the Washington Academy of Sciences*, 1919, Vol. 9, p. 217.

1

Fig. 0.1 Hickson compact group 87, about 400 million light-years away in the constellation Capri-cornus. The group shows the main morphological types of galaxies: spirals and ellipticals. Viewed from the same distance, the Milky Way would look like the tilted spiral at the top of this image. Credit: Gemini Observatory.

Turkey, it is hard to imagine that thirteen centuries ago a battle about images shook the ancient city to its societal underpinnings. These were the times of the iconoclasts, those who opposed religious images and who destroyed them. Anecdotal history of Byzantium reports that the first iconoclast emperor Leo III (c. 685–741) not only closed the imperial university, but also had it burnt down along with the library and its professors. Historians have shown that the rise to political power of the iconoclast regime and oligarchy in the eastern Byzantium Empire during the eighth century triggered a decadent period for Greek science. Although the burning story is considered apocryphal, it suggests that the alliance between images, knowledge and the freedom to search for knowledge have deep roots in human intellectual history. "No icons. No Science."[4]

Images of celestial objects are unlikely objects of veneration; making and studying astronomical images have rarely been controversial. Nevertheless, the association between the night sky and images runs deep. For thousands of years, humans have associated the patterns of stars in the sky with mythological figures for mnemonics and also for expressing their awe at the mysterious celestial vault. "By connecting dots with lines and parts with wholes, relations and structures appeared. This marvel had been well known since ancient times, at least ever since constellations of bears, angels, heroes, and swans were first marked out in the sky."[5] Constellations are images for the mind, and at the same time a tool to assign order in the sky while inventing personages or beasts for dreams and imagination. The perpetuity of the use of constellations and asterisms, from Antiquity to our times, is a testimony to the enduring conceptualizing power of images.

As in other fields of scientific investigation, astronomical images have generally been considered an accepted, and even required, procedure to report on the nature of things. I will show that there can be a tortuous path between seeing something and realizing what it really is or where it is. Images are a tool of discovery, but are not necessarily self-evident. The scientific apparatus used to take or reproduce images helps us to see things in new, sometimes revolutionary, ways. Scientific images are also an effective means to share information with other scientists and the public. This is a theme that has been well studied for other sciences, for example in geology by British geologist and historian Martin J. S. Rudwick.[6]

Both the discovery process and the sharing are at the root of the creation of the compendium of images, the scientific atlas.[7] In an atlas, pictures are put into a framework. As I will demonstrate, the diversity of form and morphology of galaxies allows their images to be put side by side in an organized series. As exemplified by the works of German naturalist Ernst Haeckel (1834–1919), the juxtaposition of images does not necessarily record an evolution but a continuity or transition in physical processes or environmental effects.[8] This is a theme that we will come back to in the later parts of the book.

Let us set the scene by examining the epitome of a discovery image in nineteenth-century astronomy.

Seeing Spirals

In the late eighteenth century, astronomers William and Caroline Herschel had found "nebulae" to pepper the sky in almost all directions. German naturalist and explorer Alexander

---

[4] E. Nicolaidis, *Science and Eastern Orthodoxy: From the Greek Fathers to the Age of Globalization*, Baltimore: Johns Hopkins University Press, 2011, pp. 40–53.

[5] O. W. Nasim, *Observing by Hand: Sketching the Nebulae in the Nineteenth Century*, Chicago: University of Chicago Press, 2013, p. 155.

[6] M. J. S. Rudwick, The Emergence of a Visual Language for Geological Science 1740–1840, *History of Science*, 1976, Vol. XIV, pp. 149–195.

[7] L. Daston and P. Galison, The Image of Objectivity, *Representations*, No. 40, Special Issue: Seeing Science (Autumn, 1992), University of California Press, pp. 81–128.

[8] O. Breidbach, *Visions of Nature, The Art and Science of Ernst Haeckel*, Munich: Prestel Publications, 2006, p. 24.

von Humboldt (1769–1859) was a visionary scholar. Commenting on the nature of these "nebulae," he boldly coined the expression "island-universes," a magnificent and visually evocative concept.[9] In his view, the Milky Way was a giant system of stars that seen from a great distance would just appear like one of the many "nebulae" reported by astronomers. The concept had already been explored in 1755 by his fellow countryman, philosopher Immanuel Kant (1724–1804), who mused over the existence of other universes just like our Milky Way (Chapter 5).[10] Both men were fantasizing: neither Kant nor von Humboldt had any scientific basis for their daring concept. As per Heber Curtis's quote at the start of this introduction, it was a hypothesis explicitly put forward long before it was actually substantiated by evidence.

About a century after Kant proposed his hypothesis, an Irish gentleman was about to make a giant leap in probing the "islands" of the cosmic sea. He did it by producing stunning hand-drawn portraits of the numerous and then still mysterious foamy patches popping up across the whole sky. On a cool and clear night in the spring of 1845, there may have been some howling wolves, and certainly some squealing wheels moving the new cantankerous machine at Birr Castle, Ireland. William Parsons, the Third Earl of Rosse, was using a new giant telescope he had just built and put into operation. Parsons had been observing a "nebula" in the constellation of Canes Venatici. The sidereal cloud he was examining was already known as listing 51 in the catalogue of French comet chaser Charles Messier. Carefully looking through his powerful instrument, Parsons noticed this time something very peculiar: the "nebula" had an overall pattern that appeared like a set of winding arms, "an arrangement of curved branches, which cannot well be unreal, or accidental."[11]

Scrutinizing what we now know to be a large galaxy located at about 23 million light-years, Parsons had come across a fundamental shape of galaxies. That night he had discovered a spiral structure (Fig. 0.2). British science historian John North (1934–2008) remarked that Parsons "did not fully appreciate what he had found for some time."[12] However, when Rosse presented his drawings to the British Association for the Advancement of Science, John Herschel made a big deal of them; it must surely have excited the lord. Today, we know that spirality betrays a pattern of subtle motions generated by a dynamic instability of stars and gas clouds orbiting in the gravitational potential of large disk-shaped galaxies.

During the following 30 years, William Parsons and Birr Castle's skillful observers catalogued thousands of "nebulae." The drawings of "nebulae" executed by the Parsonstown's observers have made history (Chapter 2). Astrophotography later showed that most of their hand-made sketches or portraits were generally correct depictions of the cosmic objects they viewed through the eyepiece of the Leviathan (Chapter 3). Yet neither the Earl of Rosse nor his observers had even the slightest idea of what these objects really were. They were also unaware that, soon following in their footsteps, astronomers would enlarge the size of the universe by unimaginable proportions.

[9] A. von Humboldt, *Cosmos: A Sketch of the Physical Description of the Universe*, New York: Harper & Brothers, 1866.

[10] I. Kant, *Universal Natural History and Theory of the Heavens* (translated by Ian Johnston), Arlington: Richer Resources Publications, 2008.

[11] C. Parsons (editor), *The Scientific Papers of William Parsons, Third Earl of Rosse 1800–1867*, London, 1926; Cambridge: Cambridge University Press, 2011, p. 116.

[12] J. North, *Cosmos: An Illustrated History of Astronomy and Cosmology*, Chicago: University of Chicago Press, 2008, p. 592.

Fig. 0.2 Early sketch of spiral "nebula" Messier 51 made by William Parsons in April 1845 (compare with Fig. 2.10a). The Whirlpool Galaxy is located at about 23 million light-years. Courtesy of Wolfgang Steinicke.

A Rapidly Expanding Universe

Coming 2,000 years after Greek astronomer Aristarchus of Samos (c. 310–230 BC), Polish astronomer and mathematician Nicolaus Copernicus (1473–1543) had blown away the size of the Ptolemaic universe by proposing the heliocentric model for the solar system. Copernicus was attempting to solve the inconsistencies of the Ptolemaic model that had placed the Earth at the center of the universe. Instead, Copernicus had the Earth rotating on itself, the Moon circling it, and the tandem in revolution around the Sun along with the five other visible planets. Like Aristarchus, he positioned the fixed stars as other suns at such great distances that they appeared to us as just faint points of light. While the Ptolemaic universe had a size 10,000 times the size of the Earth, Copernicus had it at least billions of times larger.[13]

During the first decades of the twentieth century, our cosmological perspective was again dramatically overturned in the wake of the works of several European and American astronomers. Unlike Copernicus, these astronomers were not trying to solve tricky celestial mechanics issues. Working collectively and competitively Heber Curtis (1872–1942), Ernst Öpik (1893–1985), Knut Lundmark (1889–1958), Vesto Slipher (1875–1969), Harlow Shapley (1885–1972), Edwin Hubble (1889–1953) and Milton Humason (1890–1972)

---

[13] A. Van Helden, *Measuring the Universe*, Chicago: University of Chicago Press, 1985, pp. 28–40.

were trying to determine the distance to intriguing celestial 'foamy patches' (Chapter 5). In doing so, they unveiled the world of galaxies in a sequence of stunning observational findings derived mainly from images obtained with a new generation of powerful telescopes.

By bringing out the world of galaxies, modern astronomers exploded the volume of the Copernican universe by at least 10^{15} times, or one million billion times. The distances to island-universes were found to be colossal, measured in millions, even billions of light-years, millions of times further away than any of the stars we see at night. The light-year is a commonly used unit of astronomical distance, not of time as its name might suggest. It is the distance travelled by light in one year at the velocity of 299,792,458 meters per second; this is equivalent to 9.46 x 10^{12} km. The new cosmological findings congealed rapidly during the first decades of the twentieth century through a succession of other unexpected observations. It was realized that the universe was not only gigantic but also almost empty. Indeed, if one took all the matter of the observable universe and collapsed it into a pancake-shaped disk with the density of water, the thickness of this flat universe would be only one millimeter! As French mathematician and writer Blaise Pascal (1623–1662) wrote in *Pensées* of 1670, "Through space the universe grasps me and swallows me up like a speck; through thought I grasp it."

More surprises came. The biggest one: the universe is expanding (Chapter 5). Moreover, one strange property does not come up alone. During the 1930s, astronomers found out with puzzlement that the dominant form of matter in the universe is invisible. Galaxies are decoys for something that is unseen but more pervasive, which is weaving the fabric of cosmic space. Because this invisible mass does not emit or absorb light, it was named dark matter. I will show how images help unravel these surreal discoveries.

These latest developments came about very quickly, resulting from observations by inquisitive minds using a succession of ever more powerful telescopes. However, the discovery of galaxies has been an amazing odyssey. Again, images were the bonfires lighting the path.

Tools of Vision and of Measurements

The appearance of vision in living organisms and the evolution of the eye have been remarkable processes. The retina of the human eye is the surprising achievement of some billions years of life evolutionary processes. Seven hundred million years ago, animals developed the light-sensitive protein opsin, which captures light. Since then, vision has taken multiple forms and provided animals of all kinds with a most efficient advantage.[14,15] Aeons later, humans find themselves equipped with a liquid ball, a flexible soft lens, and 125 million photosensitive cells feeding a brain powerfully adapted for vision. Vision has provided us with an added capacity to adapt and survive, as well as to be creative and innovative enough to find stars, nebulae and explore the extragalactic universe. All this because we have eyes

---

[14] R. Dawkins, *The Blind Watchmaker*, New York: W. W. Norton & Company, 1996.
[15] Extinct trilobites had solid eyes made of calcite. See L. Browers, Animal Vision Evolved 700 Million Years Ago, *Scientific American Blogs*, November 20, 2012.

assisted by a powerful brain to process the optical signals and the billions of photons that continuously hit the retina.

We observe, astronomers observe. A simple statement attributed to the American baseball player Yogi Berra (1925–2015) is most relevant: "You can observe a lot by just watching." Historians of science Lorraine Daston and Peter Galison set this out in more scholarly words: "Perceptions, judgments, and, above all, values are calibrated and cemented by the incessant repetition of minute acts of seeing and paying heed."[16]

Curiously, the topic of scientific observation itself has not received that much attention from historians of science. Hence the fine collection of essays assembled by Lorraine Daston and Elizabeth Lunbeck is quite unique and most enlightening. The essays show the evolution of the meanings and practices of observing and experimenting from antiquity to the early twentieth century.[17] Learning to observe and to depict meant to acquire an ethos and a new way of seeing, as Daston explained. The methods of scientific observations as we know and understand them are relatively recent. Even during this period, they have evolved significantly, going initially from long lists and tabulations to include, with time, more details on set-ups, environments and conditions, bringing us finally to the modern-day metadata system. "By the turn of the eighteenth century, 'observation' had become an essential practice in almost all of the sciences, not just astronomy, meteorology, and medicine – and the complement and supplement of 'experiment'."[18] As I will show, the opening of the world of galaxies benefited from this maturing process and pushed the cognitive demands of observing. Images, obtaining and interpreting them, played a crucial role in the deep transformation of astronomical observing, from the time of Galileo telescopic viewing to modern-day 'machine' science.

Working Objects

As finely described and analyzed by Harvard University historian Peter Galison, scientists have been very creative at developing and using a whole arsenal of machines. In his monumental *Image and Logic*, Galison contrasts the visual and logical approaches of the scientific method. One tradition aims at producing "images of such clarity that a single picture can serve as evidence for a new entity or effect."[19] Against imaging, Galison contrasts the "logic tradition" where electronic devices, amassing signals, act as counting machines instead of picturing machines. In all fields of science, quantitative measurements are an essential basis for assembling factual data in order to identify processes, reconstruct behaviors and establish trends. The two approaches complement each other.

In the battery of tools for observing and measuring, images and the ways to record them are outstanding. "Part of the *déformation professionnelle* of scientific observers is a

---

[16] L. Daston and P. Galison, *Objectivity*, New York: Zone Books, 2007, p. 367.

[17] L. Daston and E. Lunbeck (editors), *Histories of Scientific Observation*, Chicago: University of Chicago Press, 2011.

[18] L. Daston, The Empire of Observation, 1600–1800, in *Histories of Scientific Observation*, L. Daston and E. Lunbeck (editors), Chicago: University of Chicago Press, 2011, p. 85.

[19] P. Galison, *Image and Logic: A Material Culture of Microphysics*, Chicago: University of Chicago Press, 1997, p. 19.

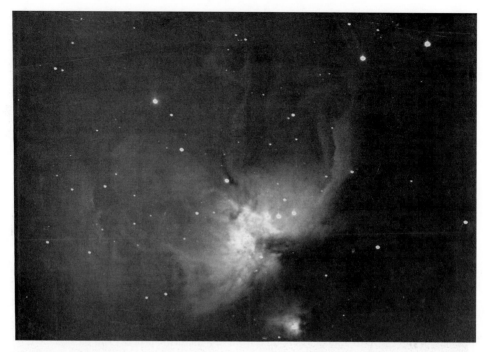

Fig. 0.3 Orion Nebula photographed with the 36-inch telescope by Andrew Common in 1883. Credit: Institute of Astronomy, University of Cambridge.

near-obsessive preoccupation with their objects of inquiry."[20] What is captured in an image can often be measured and put into graphic forms, or non-representational pictures, to summarize and convey information as effectively as possible (Chapter 4).[21] Commenting on Ernst Haeckel's experience of nature, Olaf Breidbach writes, "The very act of looking at nature was the best way of understanding it. Illustrations were not simply images of nature – they were the very embodiments of scientific knowledge. And a scientist was someone who illustrated his observations of nature in such a way as they could be shared by those not in a position to carry out discoveries for themselves."[22] The illustrations become scientific working objects.

"Working objects can be atlas images, type specimens, or laboratory processes – any manageable, communal representative of the sector of nature under investigation. No science can do without standardized working objects, for unrefined natural objects are too quickly particular to cooperate in generalizations and comparisons."[23] In astronomy, examples of working objects are archetypal objects such as the Orion Nebula (Fig. 0.3), the "great galaxy" in Andromeda (Fig. 0.4, Plate 1.1), the Omega Centauri globular cluster (Fig 0.5),

[20] L. Daston, On Scientific Observation, *ISIS*, March 2008, Vol. 99, No. 1, pp. 97–110, p. 107.
[21] E. R. Tufte, *The Visual Display of Quantitative Information*, Cheshire CT: Graphics Press, 1983.
[22] O. Breidbach, *Visions of Nature, The Art and Science of Ernst Haeckel*, Munich: Prestel, 2006, p. 20.
[23] L. Daston and P. Galison, *Objectivity*, New York: Zone Books, 2007, p. 19.

Fig. 0.4 Messier 31, the Andromeda Galaxy, imaged with the Spitzer Infrared Space Telescope. The image shows the galaxy at a wavelength of 24 microns in the mid-infrared. Credit: NASA/JPL-Caltech/Karl Gordon.

Fig. 0.5 Center of globular cluster Omega Centauri. The cluster is about the size of the full Moon in apparent size. Credit: ESO.

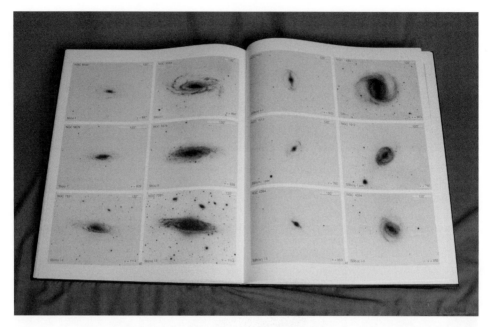

Fig. 0.6 Two pages of the NASA *Atlas of Galaxies* by Sandage and Bedke (1988). Courtesy of NASA Scientific and Technical Information Division.

the Sun and several other objects of the northern and southern celestial hemispheres. Image compendia, in particular atlases of galaxies, are other forms of working objects (Fig. 0.6). Non-representative images can also be working objects. Hertzsprung–Russell diagrams, which distribute stars of different luminosities as a function of their colours, have been powerful working objects to help understand the evolution of stars. Comparable diagrams have been created to describe and separate classes of galaxies. Working objects are the material from which concepts are developed and applied to broader classes of objects.

Images at Work

The image tradition is the center of our attention. Several groundbreaking images helped to unravel the world of galaxies, showing evidence for a new entity or effect. As a conducting thread of the book, I will present several noteworthy images. Transformational images, like Lord Rosse's first drawing of Messier 51, or the photographic plates obtained by Edwin Hubble identifying variable stars in the Great Andromeda Nebula, are not necessarily spectacular from an "aesthetic" viewpoint. Actually they are often quite bland to the unfamiliar eye. Furthermore, what appears on an image may not be at all obvious; instead it may act as evidence for something thus far unknown or poorly understood, for example the map of the distribution of galaxies in the Coma cluster and Virgo cluster of galaxies, first by William Herschel in 1785, then by Fritz Zwicky 150 years later. Only a trained eye will notice the unusual concentration and relative symmetry of these systems of galaxies.

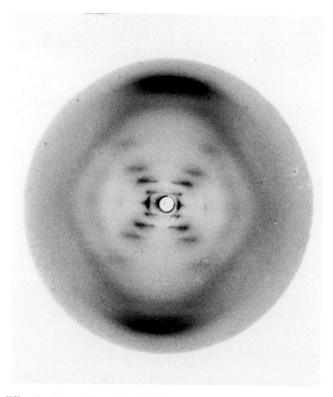

Fig. 0.7 X-ray diffraction image, Photo 51, of DNA taken by Raymond Gosling in May 1952. Gosling was working under Rosalind Franklin on the structure of DNA. Credit: King's College London.

To wrap up this introduction, I present three examples taken from molecular biology and astronomy and show how the "visual approach" helped cut through new frontiers. These examples illustrate different aspects of the cognitive challenges which images can present to the inquisitive scientist; in particular, that images are not self-evident.

The discovery of the molecule deoxyribonucleic acid (DNA) and its role in carrying genetic instructions in the reproduction and growth of living organisms is considered a momentous event of twentieth-century science. In the early 1950s, English chemist and crystallographer Rosalind Franklin (1920–1958) produced critical X-ray diffraction images of DNA, which rapidly led to the identification of the double helix structure of the molecule by James Watson, Francis Crick and Maurice Wilkins in 1953. One historical image is Photo 51 (Fig. 0.7). It is not a direct image of an object; Photo 51 shows the microscopic diffraction pattern produced by structures at the molecular level by shining X-rays on them. To get the shape of the real thing, i.e. the double helix, one needs to do a mathematical transformation of the diffraction pattern. Photo 51 was crucial to Crick and Watson for their derivation of the tilt of the helix, i.e. the angle from the perpendicular to the long axis, the distance between the chemical bases, and the length of a full turn of the helix. The

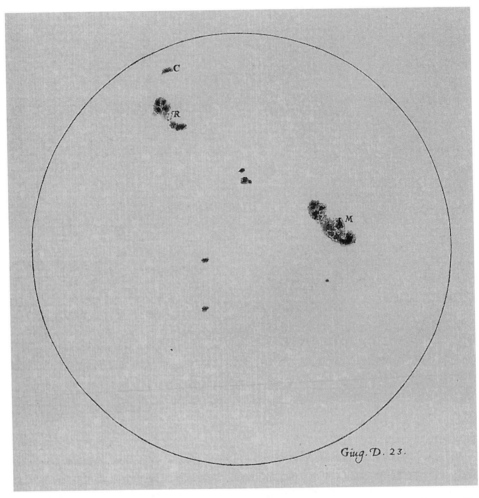

Fig. 0.8 Sunspots drawn by Galileo on 23 June, 1612. From *Istoria e Dimostrazioni Intorno alle Macchie Solarie e Loro Accidenti*. Courtesy of The Elliott and Eileen Hinkes Collection of Scientific Discovery, The Sheridan Libraries, Johns Hopkins University.

cognizant steps between seeing Photo 51 and reaching the groundbreaking conclusion of the DNA helix structure took only a few months. However, the translation of Photo 51 into modern colour-coded representations of the DNA two-base pairs is undecipherable to the unfamiliar onlooker.

The second example calls for a more recognizable cognitive operation in analyzing what is seen, but where the correct understanding of the phenomenon required centuries of investigation, with lots of speculation. Sunspots were drawn with great accuracy as soon as telescopic solar observations could be conducted in the second decade of the seventeenth century (Fig. 0.8). Early visual observations and drawings of sunspots drove Galileo Galilei

Fig. 0.9 Green pea galaxy J0925+1403 imaged with the Hubble Space Telescope. Credit: NASA, ESA.

(1564–1642) and Christopher Scheiner (1573–1650) to claim that the solar surface was not "pure and featureless." Sunspots apparently moving on the solar surface showed instead that the Sun was rotating on itself. Nevertheless, it took until the beginning of the twentieth century to explain the somber regions on the Sun not as volcanoes but as localized regions of strong magnetic fields. For almost three centuries, solar astronomers kept making superb sketches of sunspots without having a clue of what they were or what was causing them.

Finally, the recent case of the Pea Galaxy, Hanny's Voorwerp, illustrates the serendipitous power of studying images through simple visual examination. Engaged in the Galaxy Zoo project to classify a large number of galaxies from the Sloan Digital Sky Survey, the young Dutch "zooite" Hanny van Arkel came across a few small, in apparent size, weird-shaped objects (Fig. 0.9). The objects (*Voorwerp* in Dutch) came out with the dominant colour green because of strong spectral emission in the doubly ionized oxygen line at 500.7 nm due to intense star formation, hence their name. The unusual object became the archetype of a new class of galaxy. In this case, an astute non-professional observer was capable of extracting useful astrophysical information from a maze of images.

These examples show how diverse the paths from imaging to knowledge and understanding can be. In the coming chapters, we will explore the tortuous history of viewing "nebulae," imagining these foamy celestial entities and finally understanding galaxies. Instead of the few months between Photo 51 of DNA and the derivation of its double helix structure, this journey took one thousand years.

The production and use of images, as well as the fumbling with wildly diverse interpretations, are not unique to astronomy. Images have been engines of scientific exploration and discovery in many disciplines as varied as subatomic physics, geology, botany, biology and neurology. Even in mathematics, image representation is a very powerful revealer as illustrated by the vivid and spectacular Mandelbrot sets to represent fractal geometry. I will use several disciplinary examples, and I will reflect on how astronomy shares with other disciplines the discovery power of images, reminding us that there are very close links between the images, the apparatus used to obtain or reproduce them and the technologies of the time. For example, highlighting the importance of the means of reproduction for visual communication, Martin Rudwick has shown how the new techniques of aquatint, wood/steel/copper engraving and lithography made reliable reproductions possible in the 1830s and transformed geological illustrations in publications.[24]

---

[24] M. J. S. Rudwick, The emergence of a visual language for geological science 1740–1840, *History of Science*, 1976, Vol. XIV, p. 151.

Part I

Images and the Cosmos

1

Viewing Heavenly Mist

Learning to see was never, is never, will never prove effortless.
Lorraine Daston and Peter Galison[1]

Of the four thousand nebulae which have been recognized, that [Great Andromeda Nebula] which forms the subject of the present account is the only one the discovery of which preceded the invention of the telescope.
George P. Bond[2]

Can Galaxies Be Seen with the Naked Eye?

Caroline Herschel (1750–1848) is probably the most famous woman in the history of astronomy (Fig. 1.1).[3] Caroline, whom Astronomer Royal Nevil Maskelyne (1732–1811) addressed as "Dear Sister Astronomer," was indeed an outstanding woman. She was only 1.3 meters (4 ft 3 inch) tall having been struck with typhus that stopped her physical growth at the age of ten. Her physical handicap did not prevent her from becoming a fine soprano singer of Handel's oratorios, and later, an active astronomer as she embarked on the audacious ventures of her older brother William. A musician and a composer, William Herschel had moved from Hanover, Germany, to England for better opportunities in musical performance.[4] Initially venturing into astronomy as a sideline occupation, William rapidly became passionate, even obsessed, about astronomical observing and telescope making. Alexander, a younger brother, was also involved in the work of the siblings.[5]

Caroline became a skillful astronomical observer, discovering eight comets. She was outstandingly competent, highly supportive of her older brother and had her own interests in astronomy. Around year 1780, she surmised that "nebulae" were not scarce and that these objects were scattered over the whole sky. She and her brother observed and compiled "nebulae" for decades. After William's death in 1822, Caroline returned to Hanover where

---

[1] L. Daston and P. Galison, *Objectivity*, New York: Zone Books, 2007, p. 161.
[2] G. P. Bond, An account of the nebula in Andromeda, *Memoirs of the American Academy of Arts and Sciences*, New Series, 1848, Vol. 3, pp. 75–86.
[3] For a fine overview of the career and contributions of Caroline Herschel, see M. Hoskin, Caroline Herschel as observer, *Journal for the History of Astronomy*, 2005, Vol. XXXVI, pp. 373–404.
[4] W. Herschel, *Symphonies*, London Mozart Players, Matthias Bamert, Colchester: Chandos Record Ltd.
[5] M. Hoskin, *Discoverers of the Universe: William and Caroline Herschel*, Princeton: Princeton University Press, 2011, pp. 23–24.

Fig 1.1 Caroline Herschel. Credit: University of Cambridge, Institute of Astronomy.

she completed a catalogue of close to 2,500 star clusters and "nebulae." She was awarded the 1828 Gold Medal of the Astronomical Society of London (later Royal Astronomical Society) for the catalogue. British astronomy historian Michael Hoskin has written in detail about the work of the Herschels.[6] He has emphasized appropriately the contributions of Caroline and her recognition that the sky was literally peppered with nebulae: " . . . her demonstration to William that nebulae were there in abundance awaiting discovery was to prove momentous."[7] It was the spark of a grand celestial survey.

It might be challenging for us more than 230 years later to imagine how critical and upwelling was Caroline Herschel's cosmic uncovering. Anyone who has had the chance to look at a nebula or a galaxy through a small telescope will wonder how ancient observers managed to figure out these strange sidereal objects. Indeed, nebulae and galaxies observed through the eyepiece of rudimentary telescopes remain flimsy and difficult to view; it requires effort and lot of patience. Even a modern-day observer, such as Allan Sandage, could find observing work miserable and difficult at times. Today, remote-controlled

[6] M. Hoskin, *Discoverers of the Universe: William and Caroline Herschel*, Princeton: Princeton University Press, 2011.
[7] M. Hoskin, Caroline Herschel as observer, *Journal for the History of Astronomy*, 2005, Vol. XXXVI, p. 394.

telescopes with their multiple viewing monitors and comfortable observing rooms have made the nighttime work easy and convivial. It is difficult to realize that, up to quite recent times, observing was a tough nighttime job tailored for incredibly dedicated individuals. Pushing the limits of their tools and managing harsh observing conditions, the Herschels came up with impressive insights on the nature of our universe.[8] They accelerated the history of astronomy on several fronts, in particular in exploring and describing the world of "nebulae."

Observing by Viewing

Recalling an impromptu nightly excursion in the Amazonian jungle, American naturalist Edward O. Wilson writes: "The best science doesn't consist of mathematical models and experiments, as textbooks make it seem. Those come later. It springs fresh from a more primitive mode of thought, wherein the hunter's mind weaves ideas from old facts and fresh metaphors and the scrambled crazy images of things recently seen. To move forward is to concoct new patterns of thought, which in turn dictate the design of the models and experiments. Easy to say, difficult to achieve."[9] With this vivid expounding, Wilson gives us a fine summary of scientific methodology and a close description of the successive steps that pace most research ventures.

Wilson's text also illustrates in a few words the long process of humankind finally realizing that most "nebulae" in the sky were distant independent stellar systems, other giant Milky Way-like conglomerates of stars. Notable is that the process of sensing and concocting patterns depends on handling images and on the interpretation of these images by the observer. To read William Herschel's papers is to retrace a good part of the process Wilson has described.

Unaware of the nature and distances of "nebulae," the Herschels and other early observers did not know they were most often observing galaxies when looking at these celestial blurs. They were puzzled, often confused. Our predecessors considered "nebulae" either as patches of a dispersed sidereal substance, or some sort of clouds or assemblies of faint distant stars. For centuries, they could not figure out either of these alternatives. True nebulae, like the Orion Nebula, are gaseous and dusty. They concentrate along the band of the Milky Way in the thousands, and they are much less abundant than the extragalactic "nebulae" or galaxies; the latter appear to avoid the plane of the Milky Way. This is because our view through the plane of our galaxy is blocked by interstellar dust, microscopic grains of graphite and silicates that stop starlight from objects, or from being visible, further away.

As per Wilson's stumbling discovery process, generations of astronomers grappled with the reality of the nebular images and tried to discern the nature of these mysterious objects. The history of the discovery of galaxies illustrates the complicated interplay between the "mishmash" of observations and the patterns of thought as hypotheses are constructed.

---

[8] M. Hoskin, *The Construction of the Heavens: William Herschel's Cosmology*, Cambridge: Cambridge University Press, 2012.
[9] E. O. Wilson, *The Diversity of Life*, Cambridge, MA: The Belknap Press of Harvard University Press, 1992/Allen Lane, 1993.

Furthermore, new knowledge from other fields is often needed to solve puzzles in one field. The nature of true nebulae (fluorescent clouds of hot gas) came to be correctly understood only after distinguishing them from galaxies by their apparent relative receding or approaching velocities, morphologies and spectra. In the 1860s, spectra of stars and "nebulae" came to be partly understood following the work of German physicist Gustav Kirchhoff (1824–1887) and chemist Robert Bunsen (1811–1899). The full explanation of the spectra and the nature of gaseous nebulae came much later from the findings of atomic physics in the 1920s and 1930s. More on this in Chapter 5.

As Wilson points out so well, observing is fundamental to the natural sciences. The practices of observing are a key part of the scientific process. If observations are important, they are not a neutral source of information about nature. Stephen Jay Gould liked to quote Charles Darwin: "How can anyone not see that all observations must be for or against some view if it is to be of any service?"[10] Observing and viewing are complex mental and cognitive processes that need to be understood.

Observing has been a study in itself. For example, manuals on how and what to observe, or how to use a given instrument for observing, have been and continue to be important guides to scientists of all epochs. The application of common practices has been crucial in viewing objects, obtaining images, describing them and interpreting them. "The collective empiricism institutionally launched by seventeenth-century academies such as the *Academia Naturae Curiosorum* or the Royal Society of London depended on the recruitment and reciprocal calibration of observers in correspondence networks. This was most obvious in the case of weather observers, who were encouraged to standardize their instruments, hours of observation, and recording forms, but it also held for astronomy, anatomy, and natural history."[11] These practices attained maturity only after a time. Ancient scholars or naturalists did not necessarily use such standard practices. It is important to remember this to avoid the trap of historicism especially when we look at centuries of a scientific quest.

Shining Through a Horn

The Greek–Egyptian mathematician astronomer Claudius Ptolemy of Alexandria (c. 90–168) was the first to mention objects in the sky (apart from the Sun and the Moon) that were not a star, a planet nor a comet. He identified five "nebulous stars." These "nebulous stars" were the stellar open clusters h & χ Perseus (Messier 34), Praesepe (Messier 44) in Cancer, Messier 7 in Scorpius, and the star pair ν Sagittarius.[12] For example, Ptolemy described the star λ Orionis, Meissa, as "the nebulous star in the head of Orion."[13] One thousand five hundred years later, Galileo showed that Meissa was simply a pack of a few unresolved

---

[10] S. J. Gould, Why Darwin?, *The New York Review of Books*, April 14, 1996, Vol. XLIII, No. 6, p. 10.
[11] L. Daston, On scientific observation, *Isis*, March 2008, Vol. 99, No. 1, p. 102.
[12] More details in P. Kunitzsch, A Medieval Reference to the Andromeda Nebula, *The Messenger*, 1987, Vol. 49, pp. 42–43.
[13] The Almagest: in *Ptolemy's Almagest*, Translated and annotated by G. J. Toomer, Princeton: Princeton University Press, 1998, p. 382.

stars appearing to cluster around a brighter star. For Galileo and several of his contemporaries, mesmerized by telescopic viewing, all nebulous patches were expected to be resolved into stars given enough telescopic power.

While Meissa fell flat as a true nebulous object, a fuzzy patch in Andromeda, not listed by Ptolemy, turned out to be much more promising. Persian astronomer Abd al-Rahman ibn Umar al-Sufi (903–986) wrote about the "little cloud" of Andromeda in his *Book of the Constellations of the Fixed Stars* of 964; al-Sufi gave descriptions of the 48 Ptolemaic constellations, and included two drawings of each constellation, one "as seen in the sky" and one "as seen on the celestial globe."[14,15] Medieval astronomers also noticed the nebulous patch in the constellation of Andromeda. Indeed under a dark sky and a moonless night, the fuzzy object is visible to the naked eye (easily distinguishable with binoculars). Now recognized as the Andromeda Galaxy, it is one of the closest galaxies and one of the largest spirals in the local universe. Unsurprisingly, the object has occupied a special place in the development of our understanding of the world of galaxies. Apart from our own Milky Way, only four galaxies are visible to the naked eye: they are Messier 31 in Andromeda (see Plate 1.1, Fig. 0.4), Messier 33 in the Triangle, and the two Magellanic Clouds (see Plate 6.5). Nevertheless, al-Sufi's short and astute remark about a peculiar object in the northern constellation was followed by a long lapse of silence.

Early in the seventeenth century, astronomical observations were dramatically improved by the use of the telescope probably invented in Italy and improved by northern European opticians.[16] Italian astronomer Galileo Galilei and English astronomer Thomas Harriot (1560–1621) were the first astronomers known to have observed the sky with telescopes of their own making. However, observing through the primitive Galileo-type refractor was a real challenge. Initially there was questioning about the reality of features seen through the assembly of the multiple glass meniscus and within the very narrow field of view of early telescopes. It is easy to understand why a skeptic priest would raise some doubt about astronomers' assertions. However, most observers quickly admitted and concluded that what they where seeing with the magnifying optics was "true."[17]

German astronomer Simon Marius (1573–1624) was the first western individual known to have observed a "nebula" with the naked eye and with a small telescope in late 1612, and to have reported on his experience.[18] Marius had spent a few months in Prague in 1601 working and learning from Tycho Brahe, and from Johannes Kepler. We find his observing notes and comments about the Andromeda "cloud" in the preface to his *Mundus Jovialis* of 1614: "It was then visible to the naked eye, and appeared through the telescope to be composed of rays of light (*radii albicantes*), increasing in brightness as they approached the

[14] P. Kunitzsch, Sufi: Abu al-Husayn Abd al-Rahman ibn Umar al-Sufi, in *Biographical Encyclopedia of Astronomers*, T. Hockey, V. Trimble, and T. R. Williams (editors), New York: Springer, 2007.

[15] For a fine reproduction of the "cloudlike spot" in the Andromeda constellation, see Fig. 3.1 in H. Nussbaumer and L. Bieri, *Discovering the Expanding Universe*, Cambridge: Cambridge University Press, 2009, p. 28.

[16] Henry C. King, *The History of the Telescope*, Mineola: Dover Publications, Inc., 2003.

[17] M. Biagoli, *Galileo, Courtier: the Practice of Science in the Culture of Absolutism*, Chicago: University of Chicago Press, 1993, pp. 96–101.

[18] H. Nussbaumer and L. Bieri, *Discovering the Expanding Universe*, Cambridge: Cambridge University Press, 2009, pp. 27–28.

centre, which was marked by a dull, pale light, – '*in centre est lumen obtusum et pallidum.*' Its diameter was a quarter of a degree, and it resembled the light of a candle, at some distance, shining through horn."[19] This is a fine description of the object as it can be seen today through a good set of binoculars or a small telescope.

Despite the fact that Galileo scanned great portions of the sky with his telescopes, we have no evidence that he observed the Andromeda "nebula." It is likely that Galileo surmised that with greater optical power the "cloud" would resolve into stars as he had shown for Meissa. The Florentine astronomer may not have been alone in ignoring the "cloud" in Andromeda. Marius was caught in a nasty priority dispute with Galileo about who had first discovered the moons of Jupiter, resolved the Milky Way into stars, and found sunspots. The German expressed his surprise that the cloud had been noticed neither by the perspicacious Tycho Brahe (1546–1601), nor by the ancient Greek astronomer Hipparchus (c. 190– c. 120 BC), the two greatest visual astronomical observers of pre-telescopic astronomy. The enraptured Marius went overboard and claimed to have resolved the stars of several "nebulae." This did not help his credibility and, unfortunately, trash hides the gems in his contentious work.[20] Nevertheless, it is puzzling that the fuzzy patch in Andromeda was not reported more often, nor inscribed on the many celestial maps of the times.[21]

Astronomers using telescopes became more intrigued as they found more and more "nebulae" across the sky. With improved telescopes, Swiss astronomer Jean-Philippe de Chéseaux (1718–1751) and French astronomer Guillaume le Gentil (1725–1792) added a few dozen objects, some of which we now know to be star clusters that were unresolved due to the limited power of their instruments. Le Gentil also left a sketch of the Orion Nebula. A few wrote vivid descriptions of their appearance and tried to figure out what they were. English astronomer and mathematician Edmund Halley (1656–1742) was intrigued by what he called "nebulous stars" and described them as space "through which a lucid medium is diffused, that shines with its own proper lustre."

We were still in the era of positional astronomy, the branch of astronomy going back to antiquity, and to which Newtonian mechanics and the use of the telescope were then giving a new life and a new degree of certainty. Studying the clockwork of planetary orbits was serious business compared to the frailty of speculative nebular work. Sidereal astronomy was quite different, so new and so flimsy that many did not even bother with it, especially when faced with the mysterious foggy patches.

Comet Fanatics

It was soon realized that when viewed through rudimentary telescopes, comets and "nebulae" could easily be confused. The difference is that comets moved daily across the sky, changing rapidly in brightness. In contrast, "nebulae" are stationary and do not vary in

[19] G. P. Bond, An account of the nebula in Andromeda, *Memoirs of the American Academy of Arts and Sciences*, New Series, 1848, Vol. 3, pp. 75–86.
[20] S. L. Jaki, *The Milky Way, An Elusive Road for Science*, New York: Science History Publications, 1972, p. 111.
[21] D. Schultz, *The Andromeda Galaxy and the Rise of Modern Astronomy*, New York: Springer, 2012.

brightness. Comets belong to the solar system; "nebulae" do not. Still, when a presumed comet is first spotted, it might just be an existing "nebula". French astronomer Charles Messier (1730–1817) wanted to prevent false alarms. A passionate comet observer, Messier discovered 13 of them. To warn other comet chasers not to be fooled, he provided a list of "nebulae", objects that resemble comets but are not. The purpose of his catalogue was a sort of black list. Do not bother about the fixed nebulous patches: learn about their location but just ignore them!

Messier's catalogue included many objects that we now know to be galaxies. Oddly, the catalogue remains to this day a magnificent and useful compendium of deep-sky objects. Like ornithologists compiling sighting lists of birds observed, many amateur astronomers of today aim to put the whole set of Messier objects under their belt when they get their first telescope. The objects are generally relatively easy to photograph with the sophisticated equipment of today, and many amateurs produce spectacular images of Messier objects.[22]

If Messier and other comet chasers like Pierre Méchain (1744–1804), who added several objects to Messier's list, had little interest in "nebulae" – as these were just distractions and had to be skipped over – others cared about "nebulae." Investigative and broad in his approach, French astronomer Nicolas Louis de Lacaille (1713–62) paid attention. He was one of the greatest observers of the eighteenth century, a pioneer in exploring the sky of the southern hemisphere. Ahead of Messier, Lacaille had catalogued 42 nebulae and clusters of the southern sky and had proposed an informative and practical classification of "nebulae." He divided these objects into three groups: "nebulae of the first class" (or nebulae without stars), "nebulous stars in clusters," and "stars accompanied by nebulosity." Lacaille was on the right track. As noted by Swedish astronomer Peter Nilson, Lacaille's simple depiction and nomenclature happen to agree approximately with the modern division of nebulous objects into globular clusters, open clusters and gaseous nebulae.[23]

From Music to the Construction of Heavens

A huge and unexpected step in deep-sky probing came from remarkable siblings: two German musicians who had migrated to England to earn a better living and who became outstanding self-made astronomers. William and Caroline Herschel swiftly advanced to take rank among the better-known astronomers of their time.[24] Starting in 1773, William's interests shifted from music to mathematics and lenses; he became obstinately passionate about them. William started building reflecting telescopes based on the use of copper–tin alloy for the reflecting primary mirror in the early 1770s. With the help of his younger brother Alexander and teams of local artisans, William became very skillful at their

---

[22] R. Gendler, Forays into Astronomical Imaging: One Person's Experience and Perspective, *Astronomy Beat*, August 30, No. 79, 2011.

[23] Peter Nilson presents a concise summary of the history of surveying and cataloging 'nebulae' and galaxies in the *Uppsala General Catalogue of Galaxies*, Royal Society of Science of Uppsala, 1973.

[24] M. Hoskin, *Discoverers of the Universe: William and Caroline Herschel*, Princeton: Princeton University Press, 2011.

Fig. 1.2 William Herschel. Engraving by James Godby showing Herschel against background of stars in Gemini where he discovered Uranus in 1781. Credit: University of Cambridge, Institute of Astronomy.

fabrication, building and selling hundreds of telescopes over his career. He rapidly turned into a dedicated and proficient astronomical observer. In a serendipitous finding with a superb 7-ft telescope, William recognized Uranus as a true planet of the solar system in March 1781 (Fig. 1.2). King George III acknowledged the famous discovery by giving Herschel a yearly stipend and requesting him to move to Windsor as astronomer of the English court, a position created to entertain the Royal family and their visitors.

In a first telescopic program, Herschel had started observing and cataloguing stars, double stars in particular. His interest was to build a database of pairs of stars all across the sky. To assemble a large sample, William Herschel systematically scanned the sky. The goal was to measure the proper motions of these stars, i.e. stars moving with respect to each other in the celestial traffic, and more importantly to measure their parallaxes; the latter was a key for deriving the absolute distance of nearby stars by trigonometry. The parallax is a slight shift of the apparent positions of nearby stars with respect to the background of more distant ones, as the Earth swings on its orbit around the Sun during the yearly revolution. Herschel's assumption was that many of the stellar pairs were not physically bound systems, but chance alignments of a relatively nearby star along the line of sight of a more distant one, thus forming a proper set-up for determining the parallactic angle.[25] Herschel published three catalogues of double stars (1782, 1784 and 1821).

The Herschels also embarked on another observing program. With the 4-inch "sweeper" William had built for her, Caroline scanned the sky in search of new comets in the dawn sky. In her systematic sweeps, she found something odd. She noted that there were many more nebulous patches than those identified and listed by Messier and Méchain, even with additions to their earlier catalogue. Caroline was intrigued. She suggested to her brother that it might be useful to initiate a survey of "nebulae" and establish the extent of the population of these many faint diffuse objects. From 1782 to 1802, well organized and strongly motivated William, assisted by Caroline, systematically scanned the sky with a new purpose: to identify and catalogue all non-stellar objects in order to better map our great stellar system and establish the Sun's position in it. The number and distribution of "nebulae" were one aspect of Herschel's investigation. Another characteristic of "nebulae" that drew his attention was morphology or shape.

"United Luster of Millions of Stars"

Herschel spent thousands of hours observing and noting the appearances, features and structures of "nebulae," emphasizing differences as well as similarities between the various categories. He found time to write extensively about his observations. In 1785, he detailed the observed distribution of stars in the Milky Way and dealt with his preliminary findings on "nebulae."[26] Several types of "nebulae" were identified as Herschel kept finding more objects, confirming Caroline's suspicion. A few hundred non-stellar objects turned out to be star clusters or nebulosities congregating in a plane across the sky that corresponded to the band of the Milky Way. However, the great majority of "nebulae" appeared well above or below this band, numerous even in fields where stars were scarce. As we now know, these were the far more numerous and distant extragalactic "nebulae," or galaxies as they were to be called a century and a half later.

Anxious to provide details, Herschel wrote descriptions of the shapes and colours of many of the "nebulae" he found. He even speculated that the "great nebula" in Andromeda, now known as the Andromeda Galaxy, was the closest to us. Herschel also described the "nebula" in Messier 51 as a bright round nebula, surrounded by a halo or glory, and accompanied by a companion.[27] This was a surprisingly good description of the distant galaxy, taking into account the size of the telescope and the type of tarnishing metallic mirror he

---

[25] For more on the parallax program, see A. W. Hirshfeld, *Parallax: The Race to Measure the Cosmos*, New York: W. H. Freeman and Company, 2001.

[26] W. Herschel, On the Construction of the Heavens, *Philosophical Transactions of the Royal Society of London*, 1785, Vol. 75, pp. 213–266.

[27] *The Scientific Papers of Sir William Herschel*, ed. J. L. E. Dreyer, London, 1912, Vol II, p. 657. Also cited in M. Hoskin, The First Drawing of a Spiral Nebula, *Journal for the History of Astronomy*, 1982, Vol. 13, p. 97.

was using at the time. For the intrigued Herschel, "nebulae" remained a mysterious set of cosmic objects.

A fundamental barrier obstructed Herschel and all his contemporaries: they had no idea of the distances to "nebulae." By assuming that stars were other suns, eighteenth-century astronomers could get a rough estimate of stellar distances by comparing the apparent brightness of the stars to that of the Sun; they inferred correctly that stars were light-years away. However, there was no direct (or indirect) way of getting even a rough estimate of distances for "nebulae." There was no 'nebula' nearby to compare with and to serve as a gauge.

Still, this did not deter the audacious Herschel from speculation. In his early article *Construction of the Heavens* of 1785, he mused: "... the naked eye, which, as we have before estimated, can only see the stars of the seventh magnitude so as to distinguish them; but it is nevertheless very evident that the united luster of millions of stars, such as I suppose the nebulae in Andromeda to be, will reach our sight in the shape of a very small, faint nebulosity." Theorizing further, he imagined what an observer living in a distant star cluster would see, adding: "... If the united brightness of a neighboring cluster of stars should, in a remarkable clear night, reach his sight, it will put on the appearance of a small faint, whitish, nebulous cloud, not to be perceived without the greatest attention."[28] Although correct, these ruminations were then entirely conjectural.

In the same 1785 article, William Herschel noted "that remarkable collection of many hundreds of nebulae which are to be seen in what I have called the nebulae stratum of Coma Berenices." This was the first description of a cluster of galaxies (Fig. 1.3 and 1.4).[29] The stratum is now known as the Coma cluster and Virgo cluster, two large concentrations of galaxies in the northern sky.[30] Coma is one of the largest clusters of galaxies in the universe, containing 1,000 individual galaxies, located 333 million light-years away. Herschel's comment marked the discovery of some of the largest assemblies of cosmic matter. Using the word "stratum," Herschel implied layering in the structure of the universe, as for the geological strata of the Earth, perhaps inspired by the innovative work of contemporary Scottish geologist James Hutton (1726–1797). Any of us who has looked through a telescope eyepiece can only admire how observant and clear-sighted were the Herschels when viewing the sky more than 200 years ago.

Musing over the nature of the nebulous patches as assemblies of a multitude of unresolved stars, we see Herschel at first clearly favoring the "island-universe," the magic expression later put forward by Alexander von Humboldt: these numerous and distant objects, he thought, are similar but detached from the Milky Way. "As we are used to call the appearance of the heavens, where it is surrounded with a bright zone, the Milky-Way, it may not be amiss to point out some other very remarkable Nebulae which cannot well be less, but

---

[28] W. Herschel, On the Construction of the Heavens, *Philosophical Transactions of the Royal Society of London*, 1785, Vol. 75, p. 218.

[29] Charles Messier had already noticed the exceptional concentration of "nebulae" in the Virgo constellation in his catalog of 1784.

[30] M. Hoskin, *The Construction of the Heavens: William Herschel's Cosmology*, Cambridge: Cambridge University Press, 2012, p. 52.

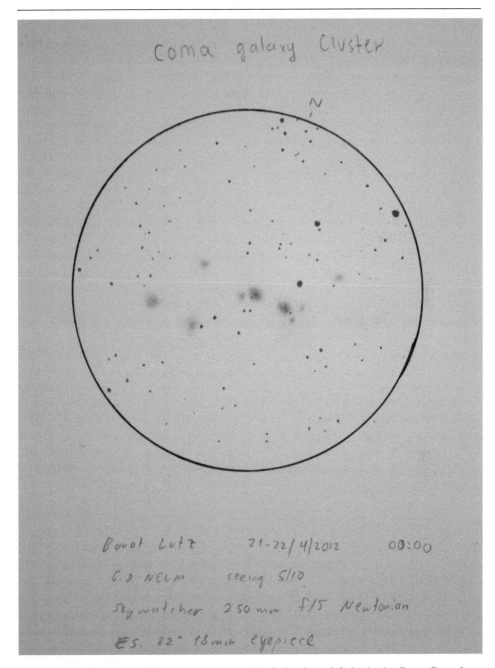

Fig. 1.3 Transformational Image: A Remarkable Collection of Galaxies in Coma Berenices. This image was viewed and sketched by amateur Michael Vlasov using a 25-cm f/5 Newtonian telescope. The view is probably similar to that observed by the Herschels. Credit: Michael Vlasov.

Fig. 1.4 Coma cluster of galaxies imaged with the 0.8-m Schulman telescope of the Mount Lemmon Sky Center. Credit: Adam Block/Mount Lemmon SkyCenter/University of Arizona.

are probably much larger than our own system . . . for which reason they may also be called milky-ways by way of distinction."[31] William Herschel held this daring opinion until 1791.

Herschel's Paradigm Shift

That year Herschel went through a paradigm conversion and became an adept of the Nebular Hypothesis. Having found a few indisputable examples of stars associated with nebulosity, Herschel flipped opinion and fixed his mind on demonstrating that all "nebulae" had to be close by. However, someone else likely stimulated his flip. In 1755, German philosopher Immanuel Kant (1724–1804) had also presented a coherent, if qualitative, proposition of "why will the middle point of every system consist of a burning body [. . .] with the sun as the central body, and the fixed stars visible to us, all things considered, mid-points of similar systems."[32]

Kant's concept was not entirely new. Inspired by the giant whirlpools of "universal fluid" imagined by René Descartes (1596–1650), Swedish mystic Emanuel Swedenborg

---

[31] M. Hoskin, *The Construction of the Heavens: William Herschel's Cosmology*, Cambridge: Cambridge University Press, 2012, p. 255–256.

[32] I. Kant, *Universal Natural History and Theory of the Heavens*, Translated by Ian Johnston, Arlington: Richer Resources Publications, 2008, pp. 105–113.

(1688–1772) had proposed elements of the nebular condensation hypothesis in 1734.[33] In 1809, French mathematician Pierre Simon Laplace (1749–1827) took Kant's idea and transformed it into a rigorous mathematical proposition. Laplace proposed a cosmogonical theory that became known as the Nebular Hypothesis.[34] As described in *The System of the World*, the Sun and the solar system formed from a large cloud of gaseous material.[35] The primordial cloud had a slight rotation. It collapsed into a disk under the action of Newtonian gravitation. With the fiery Sun at the center, the planets and their satellites kept the imprint of the original cloud motion, forcing all rotational and orbital movements to be in the same direction. Years before Laplace, Herschel already felt he was on solid ground and pushed forward with Kant's appealing idea.

Having pocketed many nebulae with a central star (e.g. the planetary nebula NGC 1514), Herschel convinced himself that nebulae of such "singular appearance" were new stars and planetary systems in formation (Fig. 1.5). "Impressed with an idea that nebulae properly speaking were clusters of stars, I used to call the nebulosity of which some were composed, when it was of a certain appearance, *resolvable*; but when I perceived that additional light, so far from resolving these nebulae into stars, seemed to prove that their nebulosity was not different from what I had called milky, this conception was set aside as erroneous."[36] Herschel invoked shining fluid not necessarily associated with a star. It was indeed a rather judicious and correct statement about galactic nebulae, as we now know them to be fluorescent clouds of interstellar gas.

Consequently "nebulae" of such "singular appearance" had to be relatively small systems and located nearby. Herschel then drew an assertive but incorrect conclusion: all "nebulae" were part of the great Milky Way system, he declared, even if he allowed some systems to be at the periphery of the Milky Way. Our great system of stars had no bound, as he wrote. In Herschel's transformed view, "nebulae" all belong to the "Heavens" defined as the Milky Way, and this same Milky Way system encompasses everything in the universe, he concluded solemnly. From today's perspective, Herschel had switched to the wrong hypothesis. His meticulous observations had provided him a reasonable justification, as more observational cases were to strengthen his "local" view. Aware of Kant's speculative work, Herschel was also attracted by its strong visual inferences and its natural link to Newton's gravitational theory. Converted into a valorous champion of the Nebular Hypothesis, Herschel was to be followed by several others in the following decades.

A note on planetary nebulae is in order. Today we now know that planetary nebulae are not associated with the birth of stellar systems but are a phenomenon that happens at the opposite end of star life. They are envelopes of gas and dust that have been spat out by stars with masses between one and seven times that of the Sun. As the central nuclear fuel runs

---

[33] E. Swedenborg, *Prodromus Philosophiz Ratiocinantis de Infinito, et Causa Finali Creationis: de que Mechanismo Operationis Animae et Corporis*, 1734.

[34] S. G. Brush, *Nebulous Earth, The Origin of the Solar System and the Core of the Earth from Laplace to Jeffreys*, Cambridge: Cambridge University Press, 1996, pp. 14–36.

[35] P. S. Laplace, *The System of the World*, Vols. 1 and 2, London: Printed for Richard Phillips, 1809.

[36] W. Herschel, Astronomical Observations Relating to the Construction of the Heavens, *Philosophical Transactions of the Royal Society of London*, 1811, Vol. 101, p. 270.

Fig. 1.5 Planetary nebula NGC 1514. Located 800 light-years away, the nebula was discovered by William Herschel in 1790. The modern image on the left was taken in the visible. The image on the right was taken in the infrared with the space observatory WISE. Credit: NASA/JPL Caltech/UCLA/Digitized Sky Survey/STScI.

out, the star goes through a complex rearrangement of its internal structure. The star interior resettles by contracting and adjusting to a new equilibrium, as its outer envelope inflates to an enormous volume, the red giant stage. At some point, the outer layers of the evolved star become detached from the star and expand further in a relatively non-violent process lasting thousands of years; the star leaves behind a small central core, which becomes a white dwarf. As the dying star evolves, its bright envelope becomes visible as a planetary nebula excited by the hot central stellar remnant. Herschel called them "planetary" because, as viewed through his telescopes, they resembled the disk of the planet Uranus.

In closing my discussion of Herschel's work, a few remarks are in order. First, Herschel had been wise to put on hold his double-star program and to moderate his ambition to measure the parallax of nearby stars. Several of the pairs he had identified turned out to be physically bound systems, thus improper for parallax determinations. Furthermore, his telescopes suffered from poor precision and accuracy; measuring a parallax with them was impossible. He could not win the race. Indeed, it was some time later, in 1838, and with much improved instrumentation, that German astronomer Friedrich Wilhelm Bessel

Fig. 1.6 Herschel's 20-ft telescope (18.7-inch mirror). Credit: University of Cambridge, Institute of Astronomy.

(1784–1846) became one of three individuals who first measured a stellar parallax. Bessel won the race by measuring a parallax of 0.314 arcsecond for the star 61 Cygni, putting it at a distance of 10.3 light-years.[37]

Secondly, to conduct his many survey programs, Herschel used several telescopes of his making. His favorite was a 20-ft focal length telescope equipped with an 18-inch metal mirror (Fig. 1.6); it was his most versatile and used instrument. Most importantly,

---

[37] Friedrich Wilhelm Bessel, Friedrich Georg Wilhelm von Struve (1793–1864) and Thomas James Henderson (1798–1864) measured stellar parallaxes almost simultaneously. Bessel had made the most extensive measurements; he received the Gold Medal of the Royal Astronomical Society in 1840. For a fine historical description, see A. Hirshfeld, *Parallax: The Race to Measure the Cosmos*, New York: W. H. Freeman and Co., 2001.

Herschel's telescopes were innovative and better than all others. The 20-year research program of William Herschel was a major accomplishment in the history of astronomy: best telescopes, best observers and best observing programs.

Perfectly Milky, Absolutely Irresolvable

The Herschels had turned an important page of nebular research, having identified and registered close to 2,500 "nebulae" and star clusters. Their work published in the *Philosophical Transactions of the Royal Society of London* greatly surpassed in number and quality of data what had been observed before.[38] Herschel's Catalogue of One Thousand new Nebulae and Clusters of Stars, later completed and published by Caroline, became the pillar of several other research programs that followed over the years, first that of his son John Herschel (1792–1871). These resulted in the General Catalogue of nebulae and clusters of stars (1864) and, in 1888, the New General Catalogue (NGC) by Danish-born Irish astronomer John Louis Emil Dreyer (1852–1926): the well-known NGC is the basis of today's galaxy catalogues.[39] Successive observers contributed to these catalogues.[40] They were based on visual telescopic observations, supplemented with written notes and descriptions. The observers often included hand-drawn sketches for the more spectacular or unusual objects (Chapter 2).

Descriptive observations of nebulae continued with John Herschel. John was the son of William Herschel and Mary Baldwin whom William had married in 1788. Educated at Eton College and St John's College, Cambridge, John's training had followed a much more scholarly path than that of his father or aunt Caroline. John received solid professional training. He turned out to be a versatile scientist, at ease with natural philosophy, mathematics and chemistry as well as astronomy.

John Herschel understood the importance of images, and took great care to make his several drawings as faithful to the true appearance as possible.[41] Although he did not appear to have tried to photograph nebulae, he later made important contributions to photography as a general technique. Writing in 1826, the young Herschel described the "great nebula" in Andromeda as "optically nebulous, owing to the smallness of its constituent stars. . . . Its nebulosity is of the most perfectly milky absolutely irresolvable kind, without the slightest tendency to that separation into *flocculi* above described in the nebula in Orion, nor is there any sort of appearance of the smallest star in the centre of the nipple. This nebula is oval, very bright, and of great magnitude, and altogether a most magnificent object."[42]

[38] H. Nussbaumer and L. Bieri, *Discovering the Expanding Universe*, Cambridge: Cambridge University Press, 2009, p. 35.

[39] W. Steinicke, *Observing and Cataloguing Nebulae and Star Clusters: From Herschel to Dreyer's New General Catalogue*, Cambridge: Cambridge University Press, 2010.

[40] For a summary of catalogues of galaxies, see H. G. Corwin Jr., Galaxy Catalogues and Surveys, in *The World of Galaxies*, H. G. Corwin Jr. and L. Bottinelli (editors), New York: Springer, 1989, pp. 1–15.

[41] O. W. Nasim, *Observing by Hand: Sketching the Nebulae in the Nineteenth Century*, Chicago: University of Chicago Press, 2103, pp. 123–170.

[42] J. F. W. Herschel, Observations of Nebulae and Clusters of Stars, Made at Slough, with a Twenty-Feet Reflector, between the Years 1825 and 1833, *Philosophical Transactions of the Royal Society of London*, 1833, Vol. 123, pp. 359–505.

As better instrumentation became available, the "nebulous cloud" of Andromeda drew more attention, this time from an observer in North America, where astronomical research was just taking its first steps at a preeminent American institution. Succeeding his father William Cranch Bond (1789–1859), American astronomer George Phillips Bond (1825–1865) became the second director of Harvard College Observatory. The younger Bond was an expert in comets, discovering ten of them. Of a broad approach in his research, Bond jumped on the opportunity to use the recently installed 15-inch "great refractor" of the Cambridge Observatory to view "nebulae." He did so with mastery. I will describe his work in more detail in the next chapter.

In the midst of the nineteenth century, astronomers were battling with "scrambled crazy images of things recently seen" as the words of Edward O. Wilson remind us. Patterns were emerging, but hypotheses were flimsy and shaky, as illustrated by William Herschel's swinging views on the nature of "nebulae." However, the process of systematic observing and cataloguing of "nebulae" set the path to the building of a solid database and for more systematic surveys. New experiments and ambitious programs were set up to pursue the chase. More rigorous approaches were taken for recording images, not just in words. Let us now see how drawing and sketching helped make progress in unraveling the mystery of "nebulae."

2

Portraying Cosmic Whirlpools

It is hard to learn to paint dirt. I painted the surface in reddish tones, but it looks now too much like Mars to me. And I've tried blue and purple, but then it looks like the Antarctic. Monet never paints dirt.

Alan Bean

Still in our climate, where there is so much cloudy weather, a year's work, measured by the number of hours when nebulae can be effectively observed, is not considerable.

William Parsons, Third Earl of Rosse[1]

In becoming familiar with something, one is on the way to becoming acquainted with its nuances, peculiarities, properties and possible nature.

Omar W. Nasim[2]

Can One Draw Objects without a Clue of What They Are?

American astronaut Alan Bean was a member of the Apollo 12 mission and of the second team to land on the Moon in November 1969. When he retired from NASA in 1981, he devoted himself to painting manned missions to the lunar surface. Although photographs showed the lunar surface to be dull gray, Bean gave it impressionistic tones with colours that reproduced the Moon environment that the astronauts experienced, "a brightly lit space where heat from the Sun is palpable."[3] Bean's inspiring and dramatic paintings are a twenty-first-century expression of a prevailing way of representing nature, by drawing it and painting it. He is an artist of nature with the eye of a scientist conveying the subtleties and challenges of depicting real objects through images.

Drawing, sketching and portraying have been part of the process of scientific description and communication for centuries. "The discovery of nature is recorded not only in the words

---

[1] Third Earl of Rosse, On the Construction of Specula of Six-feet Aperture, and a selection from the Observations of Nebulae made with them, *Philosophical Transactions of the Royal Society of London*, 1861, p. 681.
[2] O. W. Nasim, *Observing by Hand: Sketching the Nebulae in the Nineteenth Century*, Chicago: University of Chicago Press, 2013, p. 36.
[3] W. L. Fox, in *Alan Bean Painting Apollo*, Washington: Smithsonian Books, 2007, p. 19.

34

of the explorers, it is also kept before us through the illustrations accompanying the history of this research, and the pictures make it much clearer to our sense than words could ever do."[4] This is what Alan Bean tries to convey to us about the nature of the lunar surface.

Artists of Nature

In previous centuries, it was common for scientists, naturalists in particular, to be proficient in drawing. A fine example was British geologist Charles Lyell (1797–1875), as shown by the drawings accompanying his influential *Principles of Geology*, published in three volumes in 1830–33.[5] A remarkable counterexample was the great Swedish botanist Carl Linnaeus (1707–1778) who wished "to express by words all marks [of a plant *genera*] just as clearly – if not more clearly – as others with their splendid drawings."[6] Several scholars described Linnaeus as a poor draughtsman. Wilfrid Blunt wrote "Matisse once said that his ambition was to draw like his little girl of five; Linnaeus achieved this effortlessly."[7] Isabelle Charmentier has given a balanced perspective of Linnaeus's abilities with respect to the role and importance of drawing in botany that can be extended to other natural sciences. If Linnaeus was a reluctant draftsman, the discipline of botany has been a rich field for legions of artists. Belgian painter and botanist Joseph Redoute (1759–1840) was one of the most remarkable and productive of them, with a total of 486 plates for *Les Liliacées*, a plant particularly hard to preserve in any other way than through botanical illustration.[8]

More established savants had official assistants to accompany them in the field and to execute sketching work under careful instructions. This practice was widespread in botany but also in other natural sciences.[9] Along with young naturalist Charles Darwin, the *Beagle* survey expedition of 1831–1836 had on board artist and draftsman Conrad Martens (1801–1878) followed by Augustus Earle (c. 1793–1838). In the summer of 1838, Swiss biologist and geologist Louis Agassiz (1807–1873) took with him six collaborators, including an artist, to explore the glaciers of the Alps; this expedition led to several finely illustrated books and articles. English painter William Hodges (1744–1797) made sublime sketches and paintings of the various locations of the Pacific visited by James Cook during the second voyage. These were adapted and used as engravings in the original publication of Cook's journal.

Atlases and field guides of birds have been important ornithological works, providing information on species, their distribution and abundances. These works made extensive use

---

[4] A. Bettex, *The Discovery of Nature*, New York: Simon and Schuster, 1965, p. 372.
[5] C. Lyell, *Principles of Geology*, New York: D. Appleton and Company, 1872, 2 volumes.
[6] Cited in S. Müller-Wille and K. Reeds, A Translation of Carl Linnaeus's Introduction to Genera Plantorum, *Studies in History and Philosophy of Science, Part C: Studies in History and Philosophy of Biological and Biomedical Sciences*, 2007, Vol. 38, No. 3, p. 568.
[7] Cited in I. Charmantier, Carl Linnaeus and the Visual Representation of Nature, *Historical Studies in the Natural Sciences*, 2011, Vol. 41, No. 4, p. 367.
[8] C. Fisher, *The Golden Age of Flowers: Botanical Illustration in the Age of Discovery 1600–1800*, London: The British Library, 2011, p. 21.
[9] I. Charmantier, Carl Linnaeus and the Visual Representation of Nature, *Historical Studies in the Natural Sciences*, 2011, Vol. 41, No. 4, p. 394.

of maps and especially of drawings, sometimes accompanied by spectacular colour plates, such works expressing a true art of ornithology.[10] French-born American ornithologist and naturalist John James Audubon (1785–1851) was an outstanding painter of birds and of their natural habitats. The seminal *Guide to the Birds* by Roger Peterson (1908–1996) published in 1934 presented hundreds of drawings of birds of superb quality. This was the first of the Peterson field guides that assisted the interested layperson in identifying birds, insects, plants and other natural phenomena. The books used plates of drawings rather than photographs of the subjects as an effective tool to highlight key markers. Atlases of birds full of faithful drawings and paintings are among the most magnificent books of natural history. English ornithologist John Gould (1804–1881) produced superb monographs of several species of birds.[11] For example, *A Monograph of the Trochilidae, or Family of Humming-Birds*, had 418 stupendous hand-coloured plates.[12] Charles Darwin referred to John Gould's work in his *On the Origin of Species*.

An interesting aspect of display in botanical and ornithological atlases was to show indigenous plants or flowers and particular insects with the birds darting or hovering near them. A fine example is *The Book of Palms* by German botanist and explorer Carl Friedrich Philipp von Martius (1794–1868).[13] The massive *Historia naturalis palmarum*, with more than 240 chromolitographs, contained drawings of hundreds of palm families found in the various continents. Vivid scenic depictions are regularly interspersed amid the magnificent plates of multiple families and sub-families of palms. "Composite fictions" are another important dimension of scientific illustration, where visual information presents all life cycles and transformations. This is nicely shown in *The Butterflies of North America, Titian Peale's Lost Manuscript*, where the caterpillar and pupa are shown sharing the stem of their food plant with the butterfly, male and female, displaying their upper and under surfaces.[14] "In today's world of image databases and laser printer, it can be hard to grasp the dedicated labor it took to craft one of these paintings, let alone thousands. A single image involved a close collaboration among plant collectors, botanists, and entire teams of artists who specialized in the various steps it took to achieve a finished illustration."[15] Not only did the whole process involve several people, sometimes of multiple expertise, it required weeks or months of patient work and, very likely, much negotiation. Nature artists were particularly adept at presenting reconstructed scenes of ancient geological times (Fig. 2.1).

As for botany and ornithology, the field of anatomy has been a discipline where sketching and drawing have been and continue to be in constant use and development. Anatomical

[10] J. Elphick, *Birds, The Art of Ornithology*, New York: Rizzoli International Publications, Inc., 2015.

[11] J. Elphick, *Birds, The Art of Ornithology*, New York: Rizzoli International Publications, Inc., 2015, pp. 226–239.

[12] J. Gould, *A Monograph of the Trochilidae, or Family of Humming-Birds*, London: Henry Sotheran & Co., 1880.

[13] C. F. P. von Martius, *Der Buch der Palmen*, Cologne: Taschen GmbH, 2016. The three-volume *Historia naturalis palmarum* was published by T. O. Weigel of Leipzig between 1823 and 1850. The whole work weighed 28 kg with book-block dimensions of 60 × 43 cm.

[14] K. Haltman (introduction of), *The Butterflies of North America, Titian Peale's Lost Manuscript*, New York: American Museum of Natural History/Abrams, 2015.

[15] D. Bleichmar, The Geography of Observation: Distance and Visibility in Eighteenth-Century Botanical Travel, in *Histories of Scientific Observation*, L. Daston and E. Lunbeck (editors), *Histories*, Chicago: University of Chicago Press, 2011, pp. 383–395.

Fig. 2.1 Etching from *Myers Lexikon* depicting the flora of the Carboniferous in Book 15, 1885. Credit: Bibliographisches Institut.

drawings, however, do not necessarily aim for exact reproduction. Sometimes an idealized representation is sought, which can better fulfil a pedagogical purpose.[16] In anatomy, this idealization has been pushed to the extreme, with the use of separate representations for various functions of the human body: blood circulation, the brain and nervous system, the skeleton, the muscles, etc. Indeed, anatomical sketches exploit the fact that hand-drawn pictures are best at highlighting a particular organ or body part; a photograph does not show the function or organ as clearly, and might instead hide its functional role. Anatomical representations are almost abstractions for the mind, while displaying separately body functions in ways that are most useful for the neophyte and the advanced learner. Atlases of anatomy are part of a system of visual displays and schematics for dissection classes, preparations for injections and preservations in wax casts or organs in jars. These highly illustrated publications have played an important role in training students, future surgeons and midwifes.[17] In Chapters 9–11, I will describe how atlases of galaxies have fulfilled a similar role for astronomers.

More broadly, several of the most famous artists of world art history have produced sublime works illustrating nature. Leonardo da Vinci (1452–1519) and Rembrandt van Rijn (1606–1669) were among the greatest illustrators of natural scenes. They gave to the art of scientific drawing its *lettres de noblesse*. Beyond their artistic value and technical achievement, sketches and drawings have played a profound epistemological role.

Finally, illustrations are the visual tool to convey the reality of what has been observed. Images allowed the test of independent verification or reproducibility by other observers. In earth sciences, drawings and illustrations were a guarantee that the researchers themselves had been in the field. "A savant had to show that he had indeed seen these features *with his own eyes*, that he had been there and studied them for himself, before he could establish any credibility or authority to pronounce on their scientific explanation."[18] This was particularly true for the geologists, volcanologists and naturalists of the nineteenth century, such as Charles Lyell, Louis Agassiz and Charles Darwin (Fig. 2.2).[19]

Drawing Celestial Objects

Astronomy has its own challenge. The technical set-up and atmospheric conditions under which astronomical observations are conducted are highly important and affect what is viewed. The faintness of most celestial objects has always required the use of the finest instruments with large collecting light power. In spite of using the best instruments, "nebulae" appeared fuzzy, featureless and colourless. The temperamental nature of the atmosphere easily blurred or extinguished the images seen through the eyepiece. If astronomical

---

[16] L. Daston and P. Galison, *Objectivity*, New York: Zone Books, 2007, pp. 69–82.
[17] There is an extensive discussion of atlases of anatomy in L. Daston and P. Galison, The Image of Objectivity, *Representations*, No. 40, Special Issue: Seeing Science (Autumn, 1992), Berkeley: University of California Press: pp. 81–128.
[18] M. J. S. Rudwick, *Bursting the Limits of Time: The Reconstruction of Geohistory in the Age of Revolution*, Chicago: University of Chicago Press, 2005, p. 42.
[19] C. Lyell, *Principles of Geology*, New York: D. Appleton and Company, 1872.

Plate V.

VIEW LOOKING UP THE VAL DEL BOVE, ETNA.

Fig. 2.2 *View looking up the Val del Bove, Etna*. In November, 1828, Charles Lyell made the drawing on which the engraving is based. From Lyell (1830–1833), *Principles of Geology*. Courtesy of Bibliothèque, Université Laval.

sketching followed a rather conservative development, it was considered for a long time far superior to what early photography could deliver.

In the pre-telescopic age, drawings of celestial objects dealt mainly with bright objects such as the Moon, comets, meteor showers or various asterisms. On a grand scale of visual representation, several splendid atlases of the sky with stars superimposed on the mythological figures of constellations were produced: for example, the superb *Uranometria* of the German lawyer and celestial cartographer Johannes Bayer (1572–1625) in 1603. Danish astronomer Tycho Brahe (1546–1601) ran his own press at Uraniborg and he illustrated his books very well. Brahe's famous figure of the *stella nova* depicting the great supernova of 1572 in the constellation of Cassiopeia continues to be used by modern astronomers.[20] In *Sidereus Nuncius* of 1610, Galileo Galilei presented several drawings of asterisms, or remarkable groupings of stars. Notable is Galileo's sketch of stars in the belt and sword of the Orion constellation, where he distinguished stars visible with the naked eye and those seen only with the aid of his telescope.

Indeed, with the advent of the telescope, things changed dramatically by allowing views of a multitude of new objects. Sketching became essential to report and communicate the "new worlds." Words and written descriptors were not sufficient. As reproducibility by different observers was an important component of the verification process, drawing became

[20] Tycho Brahe, *Astronomae Instauratae Progymnasmata*, Frankfurt, 1610.

an essential tool.[21] However, the faintness of many of the objects under study and the turbulent atmosphere remained continuous challenges.

Studied with the telescope, the Moon was an instant favorite for sketching and graphic
reporting. Within months of each other, English astronomer Thomas Harriot and Galileo
made the first known drawings of the Moon as viewed through low-magnification telescopes. Harriot is considered to be the first to have observed the Moon through a telescope
as per his drawings executed on 26 July 1609, around 0900 hrs. As the number of telescopes
and users grew, our natural satellite was mapped in its entirety during the first few decades
of the seventeenth century.

Mapping and drawing of the Moon culminated with the publication of *Selenographia,
sive Lunae description,* in 1647; this supreme work was the milestone achievement of
Danzig mayor and master brewer Johannes Hevelius (1611–1687). For this formative work,
Hevelius has been called the founder of lunar topography. With telescopes improving over
the next two centuries, drawings of the Moon continued to be valued. Lunar drawings even
became the indicator of the power and quality of any new instrument. A climax was reached
when Scottish astronomer and engineer James Hall Nasmyth (1808–1890) presented his
drawings of the Moon "to educate the eye" at the 1851 Great Exhibition in London where
he received a gold medal.[22]

In 1659, Dutch astronomer and physicist Christiaan Huygens (1629–1695) published his
highly illustrated book, *Systema Saturnium*. In this work, Huygens summarized his research
on Saturn, its system of rings and its satellite, Titan, which he had discovered in 1655. The
sketching work was very well presented and emphasized Huygens' interpretation of the
nature of the rings. The book became the definitive work about the saturnian world, in part
because of the extent and quality of its illustrations.

Drawing the Dazzling Sun

The Sun is bright and, for early observers, its surface showed interesting features such as
sunspots and faculae (see Fig. 0.8). As these features appeared and dissolved with time,
tracking their changes became the driver for daily sketching of the solar surface. Harriot,
Galileo, German Jesuit astronomer Christopher Scheiner (1573–1650), Frisian/German
astronomer Johannes Fabricius (1587–1616) and several observers took turns at mapping
the solar surface and its transient sunspots. Astonishingly, early observers used no eye pro-
tection of any sort. Quickly peeping at the blinding image of the Sun, they got glimpses of
its dazzling surface. Thankfully, the much safer and more reliable technique of projecting
the solar image, introduced by Scheiner, was adopted.

The discussion of the nature of sunspots led to an interesting and controversial exchange
between Galileo and Scheiner. Galileo was promoting an explanation that sunspots were

[21] S. Schaffer, On Astronomical Drawing, in *Picturing Science, Producing Art*, C. A. Jones and P. Galison (editors), New York:
      Routledge, 1998, pp. 441–474.
[22] J. North, *Cosmos: An Illustrated History of Astronomy and Cosmology*, Chicago: University of Chicago Press, 2008, p. 489.

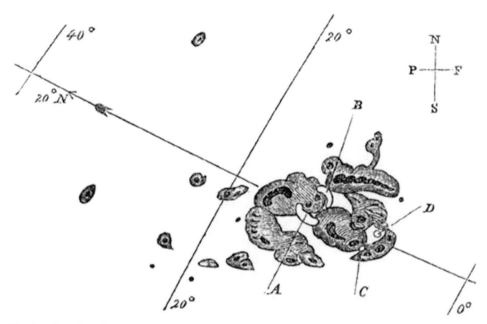

Fig. 2.3 Drawing of the large 1859 sunspot group. The white light flare is indicated as A, B, C and D, features that evolved over minutes but were caught by astronomer Richard Carrington. From Carrington (1859), *Monthly Notices of the Royal Astronomical Society.*

truly part of the surface of the Sun. Scheiner, anonymously hiding under the pseudonym Apelles, claimed for some time that they were satellites around the Sun. The observations and sketches of the Sun were the main themes of the first astronomy book on the Sun published in Italian and Latin in 1613 (see Fig. 0.8).[23] The book presented numerous sketches of the solar surface; for example, the evolution of the solar surface and its spectacular display of sunspots can be followed over several weeks of the month of June 1612 (Fig. 0.8). The features sketched are very familiar to a modern observer of the Sun.[24]

Some of the historical solar drawings have been used to reconstruct the historical record of solar activity. Interesting or new solar phenomena were discovered and recorded with the aid of drawings. German astronomer Wilhelm Tempel (1821–1889) made a drawing of the solar corona during the eclipse of 18 July 1860. By coincidence, it captured a coronal mass ejection in progress, a phenomenon that we recognize today as most important for space weather and solar–terrestrial interactions. On 1 September 1859, British astronomer Richard Carrington observed an intense solar flare in white light; the rare event lasted only five minutes, but he made an accurate drawing of it (Fig. 2.3).[25] Carrington, "the last serious

[23] *On Sunspots: Galileo Galilei & Christoph Scheiner,* Translated and introduced by E. Reeves and A. van Helden, Chicago: University of Chicago Press, 2010.

[24] H. Bredekamp, *Galilei Der Künstler: Der Mond, Die Sonne, Die Hand,* Berlin: Walter De Gruyter Inc., 2009.

[25] R. Carrington, Description of a Singular Appearance Seen in the Sun on September 1, 1859, *Monthly Notices of the Royal Astronomical Society,* 1859, Vol. 20, pp. 13–15.

and most tragic of grand amateur astronomers," had discovered solar flares.[26] He accurately sketched this particular event; it was so powerful that it appeared as an unusual imprint, a sharp localized harsh brightening of the solar photosphere. He correctly related the solar event to the powerful magnetic storm that hit the Earth 16 or 17 hours later. Reporting on Carrington's observation, the Royal Astronomical Society commentator cautiously stated: "While the contemporary occurrence may deserve noting, he [Carrington] would not have it supposed that he even leans towards hastily connecting them. 'One swallow does not make a summer'." However, Carrington had recorded the right "swallow." We know that major geomagnetic storms are triggered by solar flares.

Drawing Sidereal Condensations and "Nebulae"

Until the beginning of the twentieth century, "nebulae" were a mixed bag of objects and no one made a distinction between local fluorescent gas clouds, objects members of the Milky Way, and the giant external stellar systems, the "island-universes" known today as galaxies. Even poorly resolved star clusters were called "nebulae." Drawing, sketching and portraying of "nebulae" became the vogue. However contrary to the Moon or the Sun, "nebulae" turned out to be frustratingly difficult objects to draw.

The first known (and preserved) drawing of a nebula is attributed to Giovanni Batista Hodierna (1597–1660) for his sketch of the Orion Nebula, a nearby gas cloud made fluorescent by ultraviolet light from a few very luminous embedded young stars. A self-educated mason and shoemaker, Hodierna had become a priest and an astronomer. An enthusiastic supporter of Galileo, he studied nebulous objects, noting differences between comets and "nebulae." He surmised that comets were made of terrestrial matter, while "nebulae" were made up of stars. He listed and described 40 objects that appeared nebulous to the naked eye but resolved into stars with the aid of the telescope. He used the degree of resolvability to classify "nebulae." French astronomer Guillaume Jean-Baptiste Le Gentil (1725–1792) is also attributed an early sketch of the Orion Nebula.

As we saw in the previous chapter, French astronomer Charles Messier (1730–1817) was mainly interested in comets. Nevertheless, Messier made several drawings: among his most famous are those of Messier 31/32/110 (the Andromeda Galaxy and its two companions) and of Messier 42 (the Orion Nebula). The first trio of objects are now recognized as genuine galaxies while the latter is the archetype of star-forming regions.

At the end of his 1811 summary paper on the thousands of "nebulae" and star clusters that he recorded with Caroline, William Herschel showed two plates with engraving of "nebulae," illustrating their multiple shapes and degrees of condensation (Fig. 2.4).[27] The illustrated "nebulae" are a mixed bag and are drawn as blurry patches, some with appendices. Our modern eye, trained by photography and a better knowledge of the range of

---

[26] A. Chapman, *The Victorian Amateur Astronomers, Independent Astronomical Research Astronomy in Britain 1820–1920*, Chichester: Wiley, 1998. pp. 40–41.

[27] W. Herschel, Astronomical Observations Relating to the Construction of the Heavens, *Philosophical Transactions of the Royal Society of London*, 1811, Vol. 101, pp. 269–336.

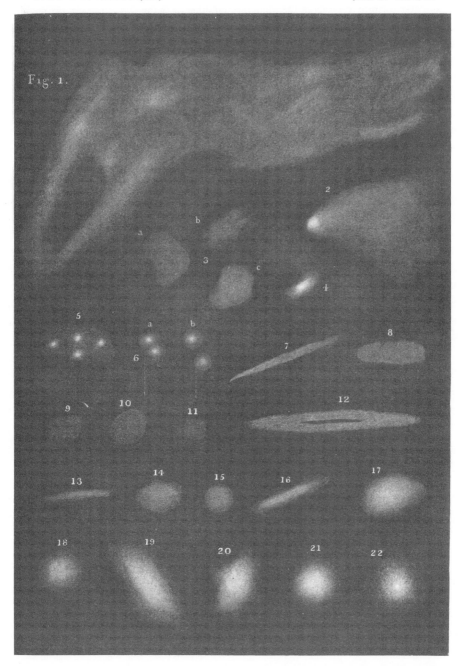

[*To face page* 480.

Fig. 2.4 Drawings of several "nebulae" by astronomer William Herschel, one of two plates he published in the *Philosophical Transactions of the Royal Society of London* (1811). Courtesy of Omar W. Nasim.

objects, does recognize the different types of "nebulae": the true nebulae as luminescent gases of star-forming regions or the planetary nebulae of dying stars, or the more regularly shaped extragalactic "nebulae," or galaxies. Herschel proposed an interpretation of the nebular shapes using the conceptual frameworks of Isaac Newton's law of gravitation and the Nebular Hypothesis. The gravitational force, he surmised, determines the shape and the degree of condensation. Herschel even speculated that comets might be the most advanced stage of nebular condensation.

The advent of several telescopes boosted interest in nebulous objects. "Gradually, nebula hunting was becoming a pursuit of its own right, although for long it was held in less esteem than comet hunting."[28] As the nineteenth century moved forward, imaging "nebulae" became an art and a science as waves of astronomers attempted to catch the elusive and diaphanous shapes. These efforts have been discussed and analyzed in depth by Omar W. Nasim in a superbly illustrated monograph presenting the works of several eminent observers: John Herschel, Samuel Hunter, Ebenezer Mason, William Lassell, Wilhelm Tempel and those of the Third Earl of Rosse's team.[29] Although astronomers drawing "nebulae" were completely ignorant or incorrectly conceived their nature, they managed to picture the objects reasonably well most of the time. The reproducibility by drawings of the same object by different observers remained questionable and some sketches are impossible to relate to the objects captured by modern photographs. Nevertheless, overall, in answer to the question of the beginning of this chapter, it can be said that it is possible to draw objects without a clue of what they are. A most dramatic example is the engraving of the great nebula in Orion by the Fourth Earl of Rosse.[30] When compared to a fine black-and-white modern photograph, the precision of the details and overall shape seen in the drawings are stunningly similar.

William Parsons and the Leviathan

The systematic use of drawings as faithful representations of "nebulae" observed through telescopes enjoyed a significant boost with the work of the Third Earl of Rosse, William Parsons (Fig. 2.5) and his team of skillful observers at Birr Castle, Ireland.

For a time a member of Parliament, Parsons ran one of the most ambitious and original scientific enterprises of the nineteenth century.[31] Well endowed through marriage with family land and fortune, Parsons leveraged these assets to realize and empower his passion for astronomy. His wife Mary Field, Countess of Rosse (1813–1885), a fine pioneer photographer herself, wholeheartedly supported her husband's astronomical projects. "Mary was an intellectual young woman, perhaps no great beauty, as she appears thin and

---

[28] J. North, *Cosmos: An Illustrated History of Astronomy and Cosmology*, Chicago: University of Chicago Press, 2008, p. 443.

[29] O. W. Nasim, *Observing by Hand: Sketching the Nebulae in the Nineteenth Century*, Chicago: University of Chicago Press, 2013.

[30] Fourth Earl of Rosse, An Account of the Observations on the Great Nebula in Orion, Made at Birr Castle, with the 3-feet and 6-feet Telescope, between 1848 and 1867, With a Drawing of the Nebula, *Philosophical Transactions of the Royal Society of London*, 1867, Vol. 158, pp. 57–73.

[31] C. Mollan, *William Parsons, 3rd Earl of Rosse: Astronomy and the Castle in Nineteenth-Century Ireland*, Manchester: Manchester University Press, 2014.

Fig. 2.5 William Parsons, 1860 portrait by Stephen Catterson Smith. Credit: University of Cambridge, Institute of Astronomy.

soberly dressed in a watercolour painted around the time of her wedding; but her serious nature suited William admirably and she concealed huge energy behind her girlish looks. Also, she was extremely rich."[32] There is evidence that Mary contributed intellectually to the lord's multiple ventures in politics, national scientific and local community affairs. Assisted by his extraordinary wife, Parsons applied his organizing talent, passion and skills in engineering to build telescopes. He adopted Herschel's proven technology of large metal-mirror telescopes and improved their design to realize larger and more ambitious instruments.[33,34]

Parsons' home, Birr Castle in Ireland, soon became a lively research center and observatory, except during the Great Famine (1846–1848), when activities were drastically reduced.

[32] Alison, Countess of Rosse, Mary, Countess of Rosse (1813–85), in *William Parsons, 3rd Earl of Rosse: Astronomy and the castle in nineteenth-century Ireland*, C. Mollan (editor), Manchester: Manchester University Press, 2014, p. 55.

[33] C. Mollan, A consummate engineer, in *William Parsons, 3rd Earl of Rosse: Astronomy and the Castle in Nineteenth-Century Ireland*, C. Mollan (editor), Manchester: Manchester University Press, 2014, pp. 159–209.

[34] C. Parsons (editor), *The Scientific Papers of William Parsons, Third Earl of Rosse 1800–1867*, Cambridge: Cambridge University Press, 2011.

Started in the late 1840s, the observatory functioned actively over more than three decades. A comprehensive paper reporting on the first years of observing with the Leviathan was published in 1861 demonstrating the power and achievements of the giant reflector.[35] In 1878, William's son, Lawrence Parsons, Fourth Earl of Rosse (1840–1908), published the results of 30 years of observing at Birr Castle.[36] Read today, these publications, together with others by Parsons and collaborators, are fascinating as they provide unique insights into the developing technology and their careful operational and observing processes. Numerous illustrations of equipment and sketches of "nebulae" with detailed descriptions supplement these publications.

Parsons' perspicacious observations had been empowered by his new giant telescope. Using first a 3-ft (mirror size) telescope based on Herschel's design, Parsons made many excellent drawings. Determined to show that a nebula could be resolved into stars and that angular resolution was related to the mirror size, Parsons built the "Leviathan," a giant telescope with a 6-ft mirror. No less important in the discovery process was the obsessive operational care of the instrument together with the innovative observing techniques Parsons and his team had put forward.

The Leviathan, with a 6-ft diameter metal mirror, was a colossal instrument for the time and remained the largest telescope until 1917. It followed the principles of Herschel's telescopes, but every aspect of its construction and material used had been greatly improved. As Herschel had not published technical details on his telescopes, Parsons had to re-invent most of the complex steps.[37] Eager to share his ideas and experiences, he described in minute detail the steps of making large speculum mirrors and assembling the complex structures for his big transit telescope. In the extensive article of 1861, Parsons aimed at giving enough details that a team of competent engineers could repeat the steps and create their own observatory.[38] However, the efforts and challenges described could frighten off even the most venturous entrepreneurs.

The 3-ft and 6-ft telescopes were strikingly impressive (Fig. 2.6). More innovative were Parsons' operational strategies, such as the careful steps taken by the team to ensure that the mirror quality remained always optimal. The speculum, a metal mirror made of a copper and tin alloy, was subject to quick tarnishing by oxidation. Therefore there were two mirrors provided for the Leviathan and two as well for the 3-ft telescope, so that one mirror might be on the polishing machine in the laboratory, while the others were in use.[39] Frequent

---

[35] Third Earl of Rosse, On the Construction of Specula of Six-feet Aperture, and a Selection from the Observations of Nebulae Made with Them, *Philosophical Transactions of the Royal Society of London*, 1861, Vol. 151, pp. 681–745.

[36] Fourth Earl of Rosse, Observations of Nebulae and Clusters of Stars Made with the Six-Foot and the Three-Foot Reflectors at Birr Castle from the Year 1848 up to the Year 1878, *Scientific Transactions of the Royal Dublin Society*, 1878, Vol. II.

[37] W. Steinicke, Birr Castle Observations of Non-Stellar Objects and the Development of Nebular Theories, in *William Parsons, 3rd Earl of Rosse: Astronomy and the Castle in Nineteenth-Century Ireland*, C. Mollan (editor), 2014, Manchester: Manchester University Press, pp. 210–270.

[38] Third Earl of Rosse, On the Construction of Specula of Six-feet Aperture, and a Selection from the Observations of Nebulae Made with Them, *Philosophical Transactions of the Royal Society of London*, 1861, pp. 681–745.

[39] G. Johnstone Stoney, in Fourth Earl of Rosse, Observations of Nebulae and Clusters of Stars Made with the Six-Foot and the Three-Foot Reflectors at Birr Castle from the Year 1848 up to the Year 1878, *Scientific Transactions of the Royal Dublin Society*, 1878, Vol. II, Appendix p. III.

Fig. 2.6 Photograph of the Leviathan. Credit: University of Chicago Photographic Archive, [apf6–01245], Special Collections Research Center, University of Chicago Library.

re-grinding and polishing were the most straightforward ways to restore reflectivity. American astronomer Ronald Buta explains that visually the Leviathan had the power of a modern 1-m-class telescope with an aluminium-coated primary glass mirror.

Attentive care was given to the operational procedures, including continuous optical testing during mirror refiguring, just as in modern optical shops. And this was quite a sizeable enterprise. A comment by George Johnstone Stoney (1826–1911), one of the Birr Castle telescope night observers, conveys a perspective on the size of the mirrors and the care taken in their manipulation. Stoney wrote: "The whole had to be lifted from the polishing machine, and transferred to a large truck, on which it was slowly dragged by twenty-five or thirty men for a distance of about a quarter of a mile to the Observatory."[40] One gets additional insight into the scale of things by noting that the Leviathan mirrors, with levers

[40] G. Johnstone Stoney, ibid., p. iii.

and carriage, weighed about seven tons. The goal of this demanding maintenance was that the mirrors had to be of a quality that their figures would match the image finesse delivered by the atmosphere on the best observing nights. And operating the Leviathan was not for the weak heart. At nighttime, the giant instrument had to be moved by two men with the aid of cables and pulleys placed between two large 18-m-high stone piers. The observer was often positioned in the dark on a high platform, tens of meters above the ground.

George Bond and the Andromeda "Nebula"

At about the same time, the young American astronomer George Bond (1825–1865) was using the new Harvard 15-inch refractor to study Messier 31. As an object of prominent interest for testing the new refractor, Bond aimed the instrument in the direction of the constellation of Andromeda. In 1848, he published a fine article with a beautiful drawing of the "nebula" and described its stars, its clusters of stars and dark bands (Fig. 2.7; compare with Plate 1.1). He reviewed his recorded historical viewings of the Andromeda "nebula": "the only one the discovery of which preceded the invention of the telescope."[41] Commenting on features already known, Bond also noticed the "sudden termination of the light on the side of the nebulae preceding in right ascension, . . . a sudden interruption of light, appearing like a harrow, dark band, . . . exterior to this, with respect to the axis, was another band or canal." He described them as "openings" and "canals." Bond was re-discovering the dust features John Herschel had reported in 1833.[42]

Bond's drawing of the Andromeda Galaxy is quite stunning even compared to modern photographs. Some remarkable features are noticeable such as the satellite galaxies Messier 32 and NGC 205, and its star-like nucleus. But most innovative for the time was Bond's excellent sketching of the two main dust lanes, or "canals," appearing against what we know now to be the nearside of the galaxy. The crispness of the observations was likely enabled by the high contrast provided by the fine optics of the new instrument. Bond gave a detailed description of the procedure to capture the fine details of the "nebula" that appeared "fifteen to twenty times" larger than the field of view of the refractor. One had to wait until Heber Doust Curtis's photographic work on spirals at Lick Observatory 60 years later to recognize that the famous dust lanes were not unique to the "great nebula" in Andromeda (Chapter 3). Bond, also a pioneer of astrophotography, passed away too soon, killed by tuberculosis at the prime age of 40.

The period between 1855 and 1876 was rich in astronomical sketching and drawing at Harvard College Observatory. Magnificent engravings of the Moon, Sun, planets and "nebulae" were produced and published by William Bond, George Bond and Joseph

---

[41] G. P. Bond, An Account of the Nebula in Andromeda, *Memoirs of the American Academy of Arts and Sciences*, New Series, 1848, Vol. 3, pp. 75–86. Some authors have suggested that Italian astronomer Giovanni Battista Hodierna would have observed Messier 33 as part of his never completed atlas *Il Cielo Stellato Diviso in 100 Mappe*.

[42] J. H. Herschel, Observations of Nebulae and Clusters of Stars, Made at Slough, with a Twenty-Feet Reflector, between the Years 1825 and 1833, *Philosophical Transactions of the Royal Society of London*, 1833, Vol. 123, pp. 359–505.

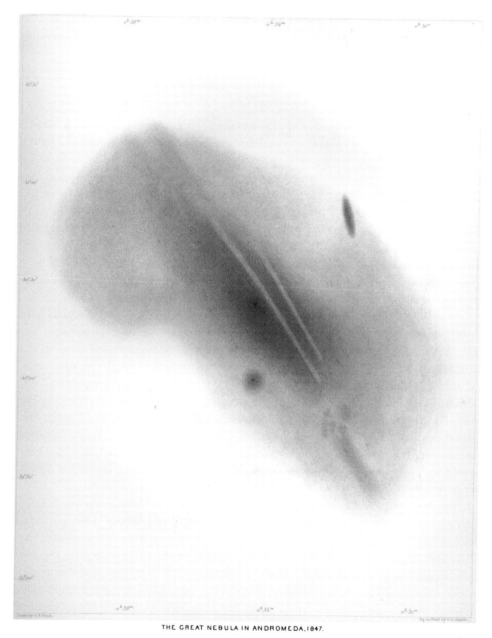

THE GREAT NEBULA IN ANDROMEDA, 1847.

Fig. 2.7 Sketch of the "great nebula" in Andromeda. From Bond (1848), *Memoirs of the American Academy of Arts and Sciences*. Courtesy of John G. Wolbach Library, Harvard College Observatory, 2016.

Winlock.[43] French artist and astronomer Étienne Léopold Trouvelot, working at Harvard Observatory, created a splendid set of sketches of astronomical objects including "nebulae."[44]

Discovering and Imaging Spirals

Following five years of construction, the Leviathan was put into operation in February 1845. The quality of early observations and findings are testimonies of the power and optical quality of this telescope and a tribute to the acuity and professionalism of the Birr Castle observers. One of the research goals of the Rosse team was to resolve "nebulae" into stars or whatever structures they might show. These drawings were generally a sketch of a single observing session. A picture or a portrait was built up over time, with pencil on paper or chalk on black cardboard. "As a collation or composition ordered and arranged by a routine procedure, therefore, what appears in a final visual result is multiple layers of different nights and days of work – a whole history of looking, discerning, and recording."[45] As mentioned above, object Messier 51 in the constellation of Canes Venatici was on the program of the first observing sessions.

It was William Parsons himself who first sketched the Messier 51 "nebula" and its spiral structure, perhaps in March or April 1845, the first portrait ever of a spiral galaxy (see Fig. 0.2).[46] John Herschel presented the early work of the Leviathan and considered it extraordinary (Fig. 2.8). The event was the Cambridge meeting of the British Association for the Advancement of Science on 19 June 1845. Herschel, who had produced a rough sketch of Messier 51 in 1833, recognized the details and the spiral structure drawn by Parsons. He praised the new drawing as a major achievement in nebular imaging. Herschel paid tribute to the Earl of Rosse and his machine with its optical quality. He stated that he himself was most familiar with the appearance of that particular "nebula" viewed in powerful telescopes, emphasizing that he could appreciate the major step forward in nebular imagery.

From Herschel's report, we get an interesting description of William Parsons' method of observing and sketching. It is worth quoting at length: "On the Nebula 25 Herschel or 61 [sic] of Messier's catalogue. Parsons exhibited to the Section what he called his work plan of this nebula and explained his method. He first laid down, by an accurate scale, the great features of the nebula as seen in his smallest telescope, which being mounted equatorially, enabled him to take accurate measurements; he then filled in the other parts, which could not be distinguished in that telescope, by the aid of the great telescope; but as the equatorial mounting of this latter was not yet complete, he could not lay these smaller portions down

---

[43] W. C. Bond, G. P. Bond and J. Winlock, Results of Observations, *Annals of the Astronomical Observatory of Harvard College*, 1876, Vol III.

[44] E. L. Trouvelot, *Astronomical Sketches Taken at the Harvard College Observatory 1878–18*. Originals are held at the Harvard University John Wolbach Library.

[45] O. W. Nasim, *Observing by Hand: Sketching the Nebulae in the Nineteenth Century*, Chicago: University of Chicago Press, 2013, p. 18.

[46] M. Hoskin, The First Drawing of a Spiral Nebula, *Journal for the History of Astronomy*, Vol. XIII, 1982, pp. 97–101.

Fig. 2.8 John Herschel, from an engraving by William Ward of a painting by H. W. Pickersgill. Credit: University of Cambridge, Institute of Astronomy.

with rigorous accuracy; yet as he had repeatedly gone over them, and verified them with much care, though by estimation, he did not think the drawing would be found to need much future correction."[47] Herschel was himself a fine artist astronomer, as testified by his beautiful hand drawing of the Large Magellanic Cloud (Fig. 2.9).[48]

Not surprisingly, M51 continued to be a prime target for the Leviathan observers. The numerous sketches made later captured all the main features of the Whirlpool Galaxy. The most faithful and dramatic portrait of the galaxy was published as a full-page plate in the seminal 1878 paper that assembled all the observing notes made over 30 years by the Leviathan observers. Messier 51 got its nickname, the "Whirlpool Galaxy," from the

[47] J. Herschel, *Report of the Fifteenth Meeting of the British Association for the Advancement of Science held at Cambridge in June 1845*, London, 1846, p. xxxvi.
[48] D. L. Block and K. C. Freeman, *Shrouds of the Night, Masks of the Milky Way and Our Awesome New View of Galaxies*, New York: Springer, 2008.

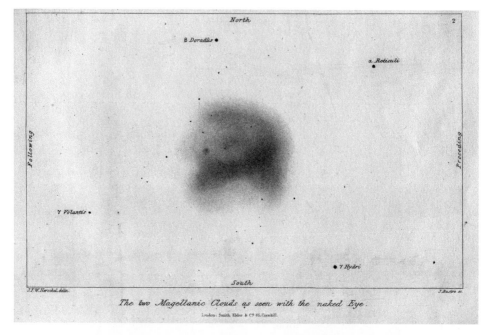

Fig. 2.9 Large Magellanic Cloud as seen with the naked eye from South Africa. Published in 1847, the superb drawing shows the elongated bar and hints of arms. Courtesy of William Cullen Library, University of the Witwatersrand, Johannesburg.

spiral structure that appeared as an indication of vortices and motions, A vivid comparison can be made by looking at the M51 portrait of 1878, based on the Leviathan telescope observations, alongside a recent image of the same galaxy produced with the Hubble Space Telescope (Fig. 2.10 a and b; see also Fig. 0.2). On the reproduction of the drawings executed at the telescope, Lawrence Parsons, William's son, explained: "Into the text have been introduced diagrams, which are rough copies of those drawn at the telescope. They have been executed by the pantagraph process employed by the Patent Type Founding Company for reproducing the weather charts for the daily newspapers."[49]

Following Messier 51, many other spirals were identified and sketched. The search for spirality superseded the drive for resolvability into stars. Drawing gained favor as the indispensable tool of recording and spirality was soon found in several other "nebulae." "Internal to the Rosse project there began a concerted effort to 'resolve' the nebulae not so much into stars but into a 'few normal forms,' the most powerful being, of course, the spiral."[50] The observers sometimes went overboard in seeing spirality where there was none, e.g. object h 311 (NGC 1514) turned out to be a planetary nebula, hence the comment of Wilhelm

[49] Fourth Earl of Rosse, Observations of Nebulae and Clusters of Stars Made with the Six-Foot and the Three-Foot Reflectors at Birr Castle from the Year 1848 up to the Year 1878, *Scientific Transactions of the Royal Dublin Society*, 1878, Vol. II, p. 4.
[50] O. W. Nasim, *Observing by Hand: Sketching the Nebulae in the Nineteenth Century*, Chicago: University of Chicago Press, 2013, p. 54.

Tempel about "spiral addiction" at Birr Castle. Nevertheless, the majority of objects identified as spirals were confirmed as such by later observations, including well known galaxies such as Messier 99 in Coma Berenices, Messier 101 in Ursa Major and Messier 33 in Triangulum. "An interesting question: how many objects are actually spiral galaxies? A modern analysis yields a number of sixty-five (out of 71) – an astonishingly large fraction and a fine proof of the quality of the Birr Castle observations."[51] Some of the non-spiral objects were galaxies of other morphological types.

The Rosse project cast its net broadly.[52] The Birr Castle program surveyed the whole sky for more nebulae and star clusters. The program added thousands of objects to those already identified and catalogued. Missing objects were searched and classified, if needed, as spurious observations or objects mistaken for their incorrect positions on the sky. Parsons' son Lawrence, the Fourth Earl of Rosse, prepared a comprehensive article of the decades of work. It was published in *The Scientific Transactions of the Royal Dublin Society.*[53] Reading the paper and the observer's notes, one comes across several mentions of spirality. Moreover, from the many drawings and notes, it is clear that the Birr Castle observers also identified several other fundamental features of galaxy morphology, such as spheroids, ellipticals, edge-on orientations or spindle shapes, barred structures, etc. The Leviathan observers also discovered multiple groupings and clusters of "nebulae."

Meticulous and reliable observations of "nebulae," many rendered in sketches and portraits, were the first aim of Parsons' empiricist enterprise. Textual notes added details on "resolvability," characteristic features, etc. Parsons' team continued to call them "nebulae," cautious not to be carried away by the ongoing dispute about the nature of these mysterious objects. It is remarkable that, even through furious ongoing debates, the "Telescope Earl" remained neutral and objective, always keeping a cautious stand on the nature of "nebulae." He never jumped on the Nebular Hypothesis bandwagon. Perhaps it was because he considered himself first and foremost an engineer, with all his efforts directed to produce the most powerful machine ever to observe the sky. He was happy to leave the detailed analysis and interpretation to others whose minds and temperaments were more at ease with speculative work. Having built the very last of the great reflectors with metal mirrors, Parsons left a stunning legacy.

A Prudent Pioneer Opening a New Era

The technological achievements of Birr Castle, with the recordings and publications, were momentous in the history of astronomy. The telescopes, the model of their operation and management, the assiduity and rigor of the observing programs and the competence of the

---

[51] W. Steinicke, Birr Castle Observations of Non-Stellar Objects and the Development of Nebular Theories, in *William Parsons, 3rd Earl of Rosse: Astronomy and the Castle in Nineteenth-Century Ireland*, C. Mollan (editor), Manchester: Manchester University Press, 2016, pp. 250–252.

[52] See O. W. Nassim, *Observing by Hand: Sketching the Nebulae in the Nineteenth Century*, Chicago: University of Chicago Press, 2013, Chapters 1 and 2.

[53] Fourth Earl of Rosse, Observations of Nebulae and Clusters of Stars Made with the Six-Foot and the Three-Foot Reflectors at Birr Castle from the Year 1848 up to the Year 1878, *Scientific Transactions of the Royal Dublin Society*, 1878, Vol. II.

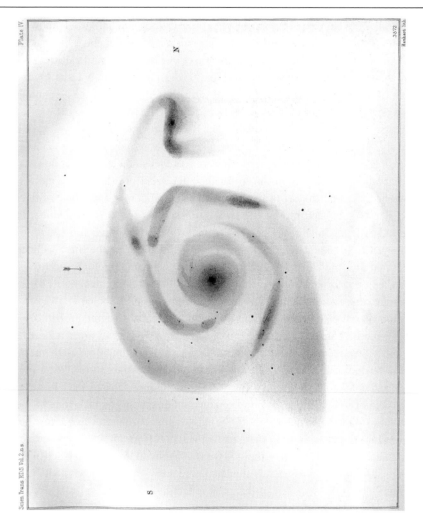

Fig. 2.10 **Transformational Image: The Spirality of Messier 51** (a) Messier 51 sketched. Portrait of the spiral galaxy (left panel) by Leviathan observers. This is a negative representation, brighter regions shown darker. From 4th Earl of Rosse (1878), *Scientific Transactions of the Royal Dublin Society*. Courtesy of The George Peabody Library, The Sheridan Libraries, Johns Hopkins University.

observers were key components for the success of the new observatory. These key elements carefully put together under William Parsons' leadership prepared the way for the great reflectors of the twentieth century (Chapter 3). Although he was not always successful, he continuously had innovation on his mind. Parsons (with the probable assistance of the Countess) attempted to photograph the Moon using the 3-ft telescope. We know that he

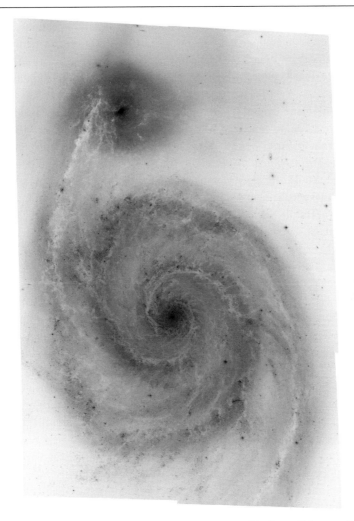

Fig. 2.10 (b) Messier 51 photographed (right panel) with the Hubble Space Telescope. This is a negative black-and-white reproduction, again with bright regions showing dark. Credit: NASA, ESA. Courtesy of Zoltan Levay (STScI).

consulted the British photographic pioneer Henry Fox Talbot (1800–1877), but that effort failed. The early photographic material was just of too low a sensitivity, and the 3-ft mount lacked the tracking advantage of the equatorial mount.[54] Parsons even tried to install a spectroscope on the 6-ft telescope. In all these efforts, the Rosse team brought observational

[54] William's son, Lawrence the 4th Earl, later transformed the original 3-ft reflector with its Herschelian mounting into a refurbished 36-inch telescope with equatorial mounting and a water-powered drive.

astronomy to another level of scientific rigour and technical know-how. "We may remark that the establishment at Birr Castle had in miniature many of the features of the modern observatory: a large telescope financed by a wealthy benefactor, a permanent skilled maintenance staff in optical and mechanical workshops, a small core of staff astronomers, and even an occasional 'guest astronomer' program with accommodation available in the Castle."[55]

Parsons and his team of observers were undeniable precursors: they opened a new astronomical field, research on galaxies. "The discoveries of spiral galaxies at Birr Castle, the excellent visual renditions made with the 72-inch *Leviathan*, and the main publications from this work in the *Philosophical Transactions* of the Royal Dublin Society for 1844, 1850, and 1861 reveal details of galaxy morphology that are beyond amazing: spiral arms, bars, rings, knots, nuclei, etc. The final summary of these observations in the *Scientific Transactions* of the Royal Dublin Society for 1879–1880 is in a way the first galaxy morphological atlas, even if galaxies are mixed with other 'nebulae'."[56]

Unfortunately, neither Parsons nor his expert collaborators had any clue of the true nature of the "nebulae." Fully aware of this, Parsons wrote: "I think we have no fair ground even for plausible conjecture; and as observations have accumulated, the subject has become, to my mind at least, more mysterious and more inapproachable."[57] It goes without saying that Parsons was right in his prudence at not letting the prejudiced mind see structure that did not exist and he warned his observing team to be most skeptical and restrained. Nevertheless, observers had managed to put together the portraits of the main nebular silhouettes (Fig. 2.11).

Notwithstanding major observational advances backed by superb images, a lot of skepticism remained with regard to the new images. Could spirality be an illusion from scattered light or an artefact of the instrument, for example circular patterns imprinted on the primary mirror during grinding and polishing? Further technical advances were needed to provide the answer. Better telescopes, not necessarily larger than the Leviathan, were a critical step.

Throughout the later part of the nineteenth century, major improvements were made to reflecting and refracting telescopes, providing aficionados of "nebulae" with more optical power. John Herschel graciously recognized the men behind this progress and the significant technical improvements, as he wrote in 1864: "Thanks to the exertion of Lord Rosse, Mr. Lassell, Messrs. Nasmyth and de la Rue in England, and Messrs. Steinheil, Foucault and Porro in Germany and France, as regards reflecting telescopes, and to those of Fraunhofer, Meez, Cauchoix, Clarke, Cook, Secretan, Ross, and Dallmeyer as regards refractors."[58] Great advances were made. Innovations included glass mirrors instead of metallic ones, optical coatings for better and durable reflection, sturdy mounting systems for accurate

---

[55] D. W. Dewhirst and M. Hoskin, The Rosse Spirals, *Journal for the History of Astronomy*, 1991, Vol. XXII, pp. 257–266.
[56] Ronald Buta, e-mail exchange with the author, 2012.
[57] C. Parsons (editor), *The Scientific Papers of William Parsons, Third Earl of Rosse 1800–1867*, Cambridge: Cambridge University Press, 2011.
[58] J. Herschel, General Catalogue of Nebulae and Clusters of Stars, *Philosophical Transactions of the Royal Society of London*, 1864, vol. 154, pp. 1–137.

Fig. 2.11 Sketches of several nebulae. Messier 51 (no. 6) and Messier 99 (no. 5) by William Parsons and Messier 31 or Andromeda Galaxy, (no. 14) by George P. Bond are easily recognized. From Alexander (1852), *On the Origin of the Forms and the Present Condition of Some of the Clusters of Stars, and Several of the Nebulae*. Courtesy of John G. Wolbach Library, Harvard College Observatory, 2016.

pointing of the telescope, and more precise clockwork for better tracking of celestial objects as the celestial sphere rotated. A new generation of telescopes and systematic observing approaches led to a re-organized exploration of the world of "nebulae." All was falling into place for the entry of astrophotography, as photographic techniques made giant leaps in sensitivity and ease of use.

3

From Celestial Snapshots to Photographing the Realm of Galaxies

I wish to mention (i) That photography has now shown itself to be capable of giving us pictures of nebulae that are superior to those made by eye and hand. (ii) That anything that can be seen by the eye with a telescope of a certain size can be photographed, and, further than this, stars that are too faint to be seen in this telescope can yet be photographed by it with sufficient exposure.

Andrew Ainslie Common[1]

A comparison of the best drawings and photographs of nebulae reveals at once the existence of considerable discrepancies between the forms depicted by methods so widely different. The actual nebulae, as photographed, have almost no resemblances to the figures.

James Keeler[2]

This series of telescopes, by revealing to all men, graphically, by means of exquisite photographs, a Universe of which the Earth, the Sun and the Milky Way are but an infinitesimal part, will bring to the world a greater Renaissance, a better Reformation, a broader science, a more inspiring education, a nobler civilization.

George Ritchey[3]

How Daguerreotype and Photography Made Astrophotography

Astronomy was one among many scientific disciplines that greatly benefited from the introduction of photography. In enabling true-to-nature representation, with the resulting "objectivity shock," photography brought about a fundamental transformation. In several ways, it meant a revolutionary change in the recording of scientific objects and phenomena.

---

[1] A. A. Common, Telescopes for Astronomical Photography, *Nature*, 1884, Vol. 31, pp. 38–40.
[2] J. Keeler, Note on a Case of Differences between Drawings and Photographs of Nebulae, *Publications of the Astronomical Society of the Pacific*, 1895, Vol. 7, pp. 279–282.
[3] G. Ritchey, The Modern Reflecting Telescope and the New Astronomical Photography, *Transactions of the Optical Society*, 1928, Vol. 29, p. 197.

The Ideal Splash and the Distractions

In order to illustrate this change, Lorraine Daston and Peter Galison most appropriately introduced the concept of "discovery shock."[4] To help us grasp the early impact of photography, these authors described the momentous experiences of British physicist Arthur Mason Worthington (1852–1916). Starting in 1875, Worthington studied the physics of splashes or the behavior of splashing liquid drops, their dynamics and the evolution of their structure, using millisecond light flashes. From the latent images left on his retina for around 10 milliseconds, Worthington depicted and drew the sequence of the evolving stages of the shape of splashing liquid drops. The sequences Worthington drew were of symmetrical and regularly shaped ribs, arms, bubbles and spouts; these certainly were pleasing to the eye. Worthington carefully produced many representative drawings that he published in professional journals over several years.

In 1894, Worthington managed to photograph the process. To his astonishment, the beautiful symmetry he thought he had faithfully captured was gone, shattered, as if it were a chimera. Instead, the photographs revealed greater irregularities than the majority of his drawings suggested. "Plunged into doubt, Worthington asked how it could have been that, for so many years, he had been depicting nothing but idealized mirages, however beautifully symmetrical."[5] Almost desperate, Worthington wrote "I have to confess that in looking over my original drawings I find records of many irregular and unsymmetrical figures, yet in compiling the history it has been inevitable that these should be rejected, if only because identical irregularities never recur. Thus the mind of the observer is filled with an ideal splash – an Auto-Splash – whose perfection may never be actually realized."[6]

A – PHOTOGRAPHING "NEBULAE," PIONEERING CELESTIAL PHOTOGRAPHY

Spiral Skeptics

The development of scientific illustrations and the use of drawings had provided scientists from the seventeenth to the nineteenth century with an efficient way to ensure a greater degree of objectivity in describing geological landscapes, clouds, animals, plants, fossils and "nebulae." By the middle of the nineteenth century, the tradition was solidly in place. But images of faint astronomical objects were flimsy. How real were the objects and features described and drawn by the observers using the Leviathan, even if it was the most powerful telescope in operation? Were the drawings just fanciful reconstructions of the mind by imaginative observers? It was not an uncommon worry, especially as several observers of "nebulae" got carried away by curvilinear shapes they believed to have seen or captured. With enthusiastic bias, some started seeing spirality where there was none, as in the well-observed Orion Nebula or Crab Nebula in the constellation of Taurus. Even the cautious George Bond stated from his observations that the Orion Nebula "may, in fact, be

---

[4] L. Daston and P. Galison, *Objectivity*, New York: Zone Books, 2007.
[5] L. Daston and P. Galison, *Objectivity*, New York: Zone Books, 2007, p. 13.
[6] A. Worthington, *The Splash of a Drop*. London: Society for Promoting Christian Knowledge, 1895, p. 74.

properly classed among the 'spiral nebulae', under the definition given by their first discoverer, William Parsons, the Third Earl of Rosse: including in the term all objects in which a curvilinear arrangement, not consisting of regular re-entering curves, may be detected."[7]

In Europe, the controversy was raging. English astronomer Richard A. Proctor (1837–1888) said that the Birr Castle drawings were the fanciful product of imagination gone astray or alluring artistic trends. German astronomer Wilhelm Tempel (1821–1889), our most vocal doubter, wrote that the spiral form did not exist among nebulae and that it was a mere creature of fantasy.[8] Were the many features of "nebulae" discovered at Birr Castle, especially the spirality that appeared to be shared by hundreds of systems, real? The skeptic Tempel was not entirely wrong in criticizing overconfident apostles of spirality, as they tend to generalize the pattern and draw more continuous structures than there really were. Photography would either confirm the reality or show these features as illusions.

Photographs versus Drawings

Photography brought an epistemological revolution. Drawing, even by a most experienced and careful observer like Worthington, could be unreliable. Worthington was extremely honest and candid in his assessment. His 1908 book summarized his research and included 197 photographs to accompany the short text.[9] While sketching, the observer consciously or unconsciously singles out features in a certain way. The photograph does not select. Consequently it records the irregularities and accidental features that the mind might consider irrelevant or distracting. This is what Worthington had experienced: the role of the self as a partial observer. The switch from recording the same experiment by visual drawing to using photography had unveiled this epistemological bias in a brutal way.

Still, despite its inherent selective mode of construction and production, most scholars agree that drawing serves an important learning function. Sketches introduce various degrees of interpretation of the object under study. With the understanding and acceptance that drawing highlights the subjective elements, the process of viewing and selecting can be seen as a teaching and learning asset. This can be employed as a pedagogical tool. As I will show later, photography did not instantaneously or completely displace drawing and sketching.

Indeed, there are cases where the science "artist" intentionally adapts the representation for educational or esthetic purposes. Of his painting *Portrait of a Crater (Second Effort)* of 1993, astronaut Alan Bean writes: "The Moon is mostly neutral gray rocks and dust with no atmosphere to soften the brilliant Sun or lighten dark shadows. To reproduce this world in paint is more science than art, so I 'key' the lunar surface neutral off-gray. That is, I paint

---

[7] G. P. Bond, On the Spiral Structure of the Great Nebula in Orion, *Monthly Notices of the Royal Astronomical Society*, 1861, Vol. 21, p. 205.

[8] O. W. Nasim, *Observing by Hand: Sketching the Nebulae in the Nineteenth Century*, Chicago: University of Chicago Press, 2013.

[9] A. M. Worthington, *A Study of Splashes*, London: Longmans, Green, and Co., 1908. The book was "dedicated to the Natural History Society of Rugby School and its former president Arthur Sidgwick in remembrance of the encouragement given to the early observations made in boyhood by my old school-friend H. F. Newal from which this study sprang."

the lunar soil slightly greenish gray, or orangish gray, and so forth. This technique opens up possibilities for hues that do no exist if a neutral gray is selected."[10] Sometimes, drawing and sketching may lead the mind the right way, possibly more effectively than photography.

Baby Steps for Astrophotography

The first successful astronomical photograph known was probably a daguerreotype of the Sun obtained by French physicist and artist Louis Daguerre (1787–1851) and astronomer François Arago (1786–1853), taken during the solar eclipse of 15 March 1839.[11] The daguerreotype was one of the first photographic processes: a latent image is recorded on a light-sensitive compound (silver bromide) and revealed by a chemical process. Early daguerreotypes required very long exposure times – several seconds for brightly lit objects. Subsequently, many researchers improved the process. A more effective process of negative-on-glass was invented in the late part of the nineteenth century and was quickly applied to astronomy. Among the first applications of astrophotography were the monitoring programs of the brighter celestial objects. For example, French astronomer Jules Janssen initiated a photographic program of the Sun at Observatoire de Meudon from 1876 to 1903. "Cinematograph" photographs of sunspots, to track their changes, were first taken around 1900. However, astronomers in general were hesitant to use the new imaging technique.

There were some practical reasons why it took some time for photography to be accepted by professionals in astronomy. Apart from lack of sensitivity, one important issue was the spreading of starry images as a blob of light on photographs, making precise astrometry much more difficult than measurements made with a micrometer based on visual observation. Furthermore, the changing image quality due to atmospheric turbulence blurred photographic images that required longer exposure times, as starlight was spread over a larger area of the photographic plates. Contrary to continuous photographic exposure, the eye can select the moments of the sharpest images. It is not surprising that several astronomers considered photographs a gimmick. For some time, most hated or ignored photography.[12] This was particularly true for the observations of the Moon and the Solar System planets. Until the invention of quick video cameras well into the twentieth century, visual observations and sketching of planet features were regarded as the best ways to capture and illustrate fine details on planets. Another damning problem with photographic emulsion was the fading of images over time due to the perishable nature of early glass negatives. Drawings could last for centuries.

Although not as bright an object as the Sun, the Moon turned out to be less challenging for early photographic work. In the winter of 1839–1840, American chemist John William Draper (1811–1892), father of astronomer Henry Draper (1837–1882), obtained

[10] A. Bean, *Painting Apollo – First Artist on Another World*, Washington DC: Smithsonian Books, 2007, p. 211.
[11] J. Lequeux, *François Arago: un savant généreux – Physique et astronomie au XIXe siècle*, Les Ulis: EDP Sciences/Paris: Observatoire de Paris, 2008, p. 424.
[12] A. Hirshfeld, *Starlight Detectives: How Astronomers, Inventors and Eccentrics Discovered the Modern Universe*, New York: Bellevue Literary Press, 2014.

a daguerreotype of the Moon through a telescope. Later, around 1850, William Bond (1789–1859), using the 15-inch refractor at Harvard College, obtained successful images of the Moon. His son, George Bond, whom we have seen observing the "great nebula" in Andromeda, pursued the improvement of photography for astronomical use. He recorded images of stars; still, his photographic plates could not reach stars of brightness fainter than those seen with the naked eye. He might have been the first astronomer to make quantitative photometric measurements on photographic plates, a truly revolutionary step.[13,14] George Bond invited American inventor and photographer John Adams Whipple to use the Harvard 15-inch and produce daguerrotypes of the Moon. The "high-tech" images, *View of the Moon*, became prize winners at the 1851 Great Exhibition in London.

Henry Draper was a pioneer of astrophotography; he photographed the Orion Nebula in 1880.[15] It showed a few fuzzy blobs, nothing to impress and stop astronomers from drawing. It was on the old continent that faster progress in astronomical photography was being made. Using his 3-ft telescope on 29 February 1883, English amateur astronomer Andrew Ainslie Common (1841–1903) also obtained a photograph of the Orion Nebula "revealing stars never seen by visual observers." He won the Royal Astronomical Society's Gold Medal for this image (Fig. 0.3). Others were catching on and progress was swift. In a major parallel effort, English astronomer and spectroscopist William Huggins (1824–1910) obtained photographic spectra (instead of images) of all bright stars between 1876 and 1886 (see Plate 5.1 for examples of modern spectra). Interestingly, the push forward was driven more by (rich) amateur astronomers, as professionals remained for a long time stubborn skeptics of the usefulness of photography for astrometry.

Dutch Prisoners Construct Maps of the Sky

Photographic techniques steadily improved in several crucial aspects, which contributed to improving their usefulness and reliability in astronomy: the production of durable photographic emulsions on dry plates; the optimization of chemical processing; better recipes for increased sensitivity; the establishment of appropriate exposure times for the various types of astronomical objects, and not least, developing robust techniques for accurately measuring the position and brightness of objects on the plates. Star catalogues were the immediate beneficiaries of the introduction of photography in astronomical work. The end result was unassailable: the positions and magnitudes of stars could be measured in much greater number and more accurately than by traditional visual astrometry.

By the late nineteenth century, the sensitivity of emulsions had improved significantly. Much fainter objects than could be recorded visually at the telescope eyepiece were being photographed reliably. The number of objects that could be measured exploded and there

---

[13] G. P. Bond, Stellar Photography, *Astronomische Nachrichten*, 1858, Vol. 49, pp. 81–100.

[14] A. Hirshfeld, *Starlight Detectives: How Astronomers, Inventors and Eccentrics Discovered the Modern Universe*, New York: Bellevue Literary Press, 2014.

[15] A. Hirshfeld (op. cit.) gives a fine overview of the development of astrophotography in the United States during the nineteenth century.

was an expressed need for a more organized effort in handling astronomical photographs. Several initiatives at photographing the sky led to an ambitious international project, the "Carte du ciel," which began in 1887. This project was to involve 20 observatories; it aimed at photographing the whole sky to a magnitude as faint as 11.5, or a couple of hundred times fainter than could be perceived by the naked eye. Well ahead of its time, the project encountered several technical difficulties and managerial problems. Soon, its promoters had to descope the venture. "When ninety-six of the Paris charts had been issued, it was estimated that the charts covering the entire sky would make a pile 120 feet high!"[16]

The Astrographic Catalogue became the more modest, but feasible, version of the project. Twenty observatories from around the world collaborated over decades, producing and measuring about 22,000 glass plates. The first such plate was taken in August 1891, at the Vatican Observatory, and the last plate at the Brussels Uccle Observatory in December 1951. The Astrographic Catalogue was largely ignored until the arrival of the Hipparcos Catalogue in 1997. Hipparcos was the European astronomical satellite that photographed 2.5 million stars for accurate star positions. One of its purposes was the accurate determination of proper motions and parallaxes. This required the use of the historical plate material for comparing changing star positions over time; the Astrographic Catalogue provided this time reference.

But let us go back to the end of the nineteenth century. Instead of aiming for the whole sky, fields of different galactic latitudes and longitudes were selected and distributed among institutions for study. A Dutch astronomer whose intent was to establish the structure of the Milky Way, the old dream of William Herschel, led a most successful initiative. Jacobus Cornelius Kapteyn (1851–1922) of Groningen University in the Netherlands managed to complete a fair fraction of a complete photographic sky survey. Between 1896 and 1900, Kapteyn measured thousands of stars recorded on photographic plates; these had been obtained by the Scottish astronomer David Gill (1843–1914) who worked at the Cape Observatory in South Africa between 1885 and 1890. Kapteyn produced a huge database of stellar positions and derived important statistical results on the distribution of stars in the Milky Way. The "Kapteyn Universe" described a lens-shaped star distribution 40,000 light-years in size, decreasing in stellar density towards its edges, and with the Sun located about 2,000 light-years from the center.

The monumental project resulted in augmenting the famous Cape Photographic Durchmusterung; this listed the positions and magnitudes of 454,875 stars between 18 degrees south in declination and the celestial north pole. There was an interesting aside to this research conducted under Kapteyn at Groningen. To accomplish the gigantic task of measuring thousands of star positions, Kapteyn requested and got the permission from the governor of the town to enroll selected male state prisoners to conduct the measurements.[17]

[16] R. G. Aitken, Dorothea Klumpke Roberts – An Appreciation, *Publications of the Astronomical Society of the Pacific*, 1942, Vol. 54, p. 219
[17] J. North, *Cosmos: An Illustrated History of Astronomy and Cosmology*, Chicago: University of Chicago Press, 2008, p. 520.

Judging from the quality of the work published, this was a successful venture of early "citizen science." One hopes that the prisoners also learned something from their unusual task.

At the same time that Kapteyn and Gill were constructing their map of the heavens, another important effort was being conducted, this time on the new continent. Harvard College Observatory entered the photographic era by initiating a systematic photographic survey of the sky (as well as spectroscopic surveys), carried out with imaging telescopes in Massachusetts, Peru and South Africa. More than 500,000 glass photographic plates were exposed between 1885 and 1993. These plates are currently being digitized and made available for reference in a modern astrometry database. At Harvard, instead of prisoners, an extraordinary female team of "human computers" were analyzing, calculating, tabulating and publishing the data extracted from the photographic plates.[18]

The Carte du ciel had gone through a near-death episode, and was only partly executed as the Astrographic Catalogue. The equivalent of the Carte du ciel was successfully resurrected half a century later, under the sponsorship of the National Geographic Society, and the Palomar Sky Survey for the northern hemisphere was thus completed in 1958. In coordination, the Science and Engineering Research Council of the United Kingdom and the European Southern Observatory sponsored the Southern Sky Atlas for the southern skies. These surveys provided an important imaging basis for identifying galaxies and clusters of galaxies. They became the reference image sources for several atlases of galaxies (Chapter 10).

The photographic plates of the two all-sky surveys were later to undergo a major revival. The scientific operation of the Hubble Space Telescope (launched in 1990) required accurate positions for the celestial objects it would be looking at. So the Space Telescope Science Institute led an enormous project: the full digitization of the thousands of survey photographic plates. The regenerated survey was released as the Digitized Sky Survey in 1994. The second-generation Guide Star Catalog was published in 2008 and it contains close to a billion objects.[19] The initial goals of the Carte du ciel were more than fulfilled 100 years later: the whole sky has been photographed and the final product is beyond what early nineteenth-century astrophotographers could have dreamt of. Yet the drive to image the whole sky has continued unabated.[20] A new generation of telescopes is being built with the sole purpose of rapidly photographing the whole sky over periods as short as a month. Leading survey telescopes are the Panoramic Survey Telescope and Rapid Response System (PanSTARRS) in Maui, Hawai'i and the Large Synoptic Survey Telescope (LSST) in Chile.

But let us return to the pioneering work of nebular photography of the nineteenth century.

---

[18] D. Sobel, *The Glass Universe, How the Ladies of the Harvard Observatory Took the Measure of the Stars*, New York: Viking, 2016.

[19] B. M. Lasker, M. G. Lattanzi, B. J. McLean, et al., The Second-Generation Guide Star Catalog: Description and Properties, *The Astronomical Journal*, 2008, Vol. 136, pp. 735–776. The GSC2.3 catalogue contains the astrometry, photometry, and classification for 945,592,683 objects.

[20] The Sloan Digital Sky Survey, or SDSS, was a photographic and spectroscopic survey conducted with a wide-field 2.5-m telescope in New Mexico, USA.

Roberts' System Engineering Approach

Because "nebulae" were faint, they proved to be very challenging targets for celestial photography. When he retired in 1888 at the age of 59, Isaac Roberts (1829–1904), a wealthy Welsh engineer and businessman, decided to devote his time to his passion for science, especially astronomy and geology. A dedicated amateur astronomer, and self-educated, Isaac Roberts rushed into nebular photography. He pushed astrophotography to new levels of performance by introducing new observing techniques and modifying the traditional telescope set-up, following an observing arrangement favored by a fellow-countryman.

Isaac Roberts, like Parsons and several others, belonged to an unusual class of individuals who conducted independent astronomical research. British geologist and historian Martin J. S. Rudwick writes about the awakening of "gentlemanly specialists" in the world of early nineteenth-century geology.[21] Astronomy, alongside geology, was an area of interest for curious minds. For British science historian Allan Chapman, "The independent tradition of astronomical research came into being, then, through a combination of intellectual curiosity, independent money, and public-spiritedness, for if you were Irish, Scottish, Welsh or English in 1840, and you wanted to attempt to 'fathom' the universe, it was no use to get into the good books of a king, a publicly funded academic body or an all-powerful minister of state in pursuit of patronage . . ."[22] Chapman described these individuals as "grand amateurs."[23] And as history showed, grand amateurs, including Parsons, were amazingly visionary and turned out to be quite successful. In particular, the Rosse program and its engineering approach served as a model for the Grubb–Parsons firm owned and led by Thomas and Howard Grubb, the great Victorian telescope makers.[24]

Particularly innovative in telescope design was William Lassell (1799–1880), a successful merchant and brewer of Liverpool. As a true grand amateur, Lassell had already demonstrated that an equatorial mount, which kept the telescope fixed in declination, with one single movement around a polar axis, was to be favored for making continuous observations. As a friend of Parsons and a frequent visitor to Birr Castle, Lassel understood the limit of the Leviathan mount. Lassel's equatorial set-up allowed the object being observed to be kept in the center of the field of view. It eliminated repeated manual re-centering and enabled uninterrupted observations for sketching nebulae or any other objects.[25]

Roberts followed in these footsteps. He understood the impact and potential of the long exposures that were allowed by dry photographic plates. The new emulsions made it possible to reach objects much fainter than anything the eye could see through the telescope. To

[21] M. J. S. Rudwick, *The Great Devonian Controversy: The Shaping of Scientific Knowledge among Gentlemanly Specialists*, Chicago: University of Chicago Press, 1985.

[22] A. Chapman, William Parsons and the Irish Nineteenth-Century Tradition of Independent Astronomical Research, in *William Parsons, 3rd Earl of Rosse: Astronomy and the Castle In Nineteenth-Century Ireland*, C. Mollan (editor), Manchester: Manchester University Press, 2014, pp. 271–297.

[23] A. Chapman, *The Victorian Amateur Astronomy: Independent Astronomical Research in Britain*, Hoboken: Wiley, 1999.

[24] I. S. Glass, *Victorian Telescope Makers, The Lives and Letters of Thomas and Howard Grubb*, Bristol: Institute of Physics, 1997.

[25] O. W. Nasim, *Observing by Hand: Sketching the Nebulae in the Nineteenth Century*, Chicago: University of Chicago Press, 2013, p. 190.

Fig. 3.1 Isaac Roberts' 20-inch reflector counter-balanced with the 7-inch Cook refractor. From Isaac Roberts (1893), *A Selection of Photographs of Stars, Star-Clusters and Nebulae*, Courtesy of John G. Wolbach Library, Harvard College Observatory. 2016.

get the full benefits of a long exposure time, Roberts improved telescope systems for accurate tracking during the several hours required to keep a given object right at the same point in the focal plane of the slowly moving telescope. He first adopted the equatorial mount design that Lassel had successfully used for operating relatively large reflectors. Roberts also introduced another critical but simple innovation.

If a telescope is being used to photograph, one cannot look into it at the same time. Despite good clock drives developed in the last decades of the nineteenth century, no tracking system was precise enough to keep an object perfectly centered over a 90-minute exposure time; fine adjustments had to be made more or less regularly. Roberts introduced the "piggyback" guide scope to follow a reference star; he mounted this second telescope in place of the traditional counterweight of equatorial mounts (Fig. 3.1). Using the 7-inch Cooke guide refractor on the same mount as the telescope used for the photography, he could keep an appropriately chosen star exactly on the reference crosshair of the eyepiece. Manually correcting any slight deviation in the tracking motion of the telescope, it was easy to compensate completely and precisely for the rotation of the Earth and for disturbances triggered by wind squalls or hiccups of the telescope clock drive. These adjustments allowed the observer to keep the camera aimed accurately during exposures lasting hours. This was crucially important for deep photography that was capable of capturing low-surface-brightness "nebulae."

All this required painstaking efforts but led to many fine photographs. After experimenting with the 7-inch imaging refractor, Roberts purchased a 50-cm reflector telescope, equipped with a very accurate clock drive, from Irish optical designer Howard Grubb (1844–1931). The camera set held 102 × 102 mm silver-on-glass photographic plates that allowed a 2 × 2 degree field of view at the celestial equator. The equipment was the heart of Roberts' private Crowborough Observatory in Sussex, England, making his home-based facility probably the first one fully dedicated to astrophotography. Robert's set-up was a marvel of innovations. His system engineering approach to telescope design and operation was a step toward the modern telescope. It was the demonstration of fine work from a true grand amateur.

Saving the Magic Half a Second

The Birr Castle observers have described eloquently how the sharpness of images seen through the telescopes changed under varying atmospheric conditions. Leviathan observer George Johnstone Stoney (1826–1911) commented that bright stars "are usually seen as balls of light, like small peas, violently boiling in consequence of atmospheric disturbance."[26] Astronomers call this boiling "seeing"; it degrades image quality, and is one of the reasons for putting telescopes in space. Moments of exquisite atmospheric stability, when for a fraction of a second the atmosphere would freeze, were rare and precious. These were the magic "half a second" instants that had to be seized when drawing or photographing.

The question in the mind of everyone who had observed through the eyepiece of a large telescope was then this: could the judgement of Birr Castle observers glimpsing faint diffuse blobs have been affected similarly to Worthington's biased perspective in studying splashing liquid drops? Were the details of "nebulae" seen and drawn as observed through the powerful 3-ft and 6-ft telescopes, but through the "boiling" atmosphere, just the "fancy of the observer" as Wilhelm Tempel claimed. Obviously bothered, Tempel had written most critically about the performance of the Leviathan, on the basis that he could not see in his 11-inch refractor the details of the "nebulae" drawn by the observers of the Leviathan. The glaring divergence between drawings of the same object particularly distressed Tempel and

---

[26] G. Johnstone Stoney, in Fourth Earl of Rosse, Observations of Nebulae and Clusters of Stars Made with the Six-Foot and the Three-Foot Reflectors at Birr Castle from the Year 1848 up to the Year 1878, *Scientific Transactions of the Royal Dublin Society*, 1878, Vol. II, p. iv.

reinforced his skepticism till the end. Joining the skeptical chorus, English science popularizer Richard Proctor (1837–1888) expressed very strong criticism about the Leviathan image quality and performance in the *Frazer's Magazine* of December 1869, and spread false rumors.[27] The patient and meticulous work of the Birr Castle observers was on the chopping board and the prognosis was quite bleak.

The Andromeda "Nebula": A Masterpiece Image

Isaac Roberts' photographs held the answers. Roberts had started photographing Messier 31, the "great nebula" in Andromeda, in the fall of 1888. Briefly commenting that the "nebula" "is now for the first time seen in an intelligent form," he was quickly dragged into a speculative discussion about the possibility of this being a new solar system that was in the process of condensing from a nebula with a central sun now seen in the midst of nebulous matter.[28] Roberts went back to the "great nebula" and his masterpiece was his 4-hour photograph of Messier 31 taken during the night of 29 December 1888 (Fig. 3.2).[29] The result must have been a relief to many: the beautiful photographic image showed a spiral structure, which was conspicuous despite the fact that the plane of the galaxy is tilted significantly into the line of sight. The very short paragraph in the article where Roberts reported his photographic results on the "nebula in Andromeda" is almost an anti-climax.[30] Roberts was actually much more voluble in his description of the photograph of the Pleiades. In March and April of 1889, 44 years after Parsons' seminal discovery of spirality, Roberts photographed Messier 51, the Whirlpool Galaxy. The image displayed magnificent spiral arms.[31]

Spirality had been found in several other "nebulae" by Birr Castle observers. Roberts now had the technique to freeze spirality on photographic plates, a much more neutral medium than drawing from visual observations through variable atmospheric conditions. Contrary to Worthington's frustrating results, photographs of "nebulae" confirmed the spirality of most of the objects which the Birr Castle team claimed to have identified intriguing patterns for. Photographs actually reinforced the regularity and symmetry of many features captured and sketched by visual observers.

The confident Roberts made regular presentations of his photographs at the meetings of the Royal Astronomical Society. Less restrained than Parsons in commenting on the nature of "nebulae," Roberts went astray in the physical interpretation of what he had photographed. He explained spiral structure as a pattern that could be the resulting material from a collision between two stars, or two "nebulae" of counter-moving streams of meteoric

[27] R. A. Proctor, The Rosse Telescope Set to New Work, *Frazer's Magazine for Town and Country*, 1869, Vol. LXXX, pp. 754–760.

[28] I. Roberts, Photographs of the Nebulae M31, h44, and h51 Andromedae, and M27 Vulpeculae, *Monthly Notices of the Royal Astronomical Society*, 1888, Vol. 49, p. 65.

[29] Roberts also used a 5 inch Cooke portrait lens that covered over 10 x 10 degrees and produced some stunning photographs.

[30] I. Roberts, Photographs of the Nebulae in the Pleiades and in Andromeda, *Monthly Notices of the Royal Astronomical Society*, 1889, Vol. 49, p. 121.

[31] I. Roberts, *A Selection of Photographs of Stars, Star-clusters and Nebulae*, London: Universal Press, 1893, Vol. 1, Plate 30, pp. 85–86.

Fig. 3.2 **Transformational Image: The First Known Photograph of a Galaxy – Messier 31 – Showing the Structure of the "Great Nebula" in Andromeda.** It was obtained by Isaac Roberts on December 29, 1888. From Isaac Roberts (1893), *A Selection of Photographs of Stars, Star-clusters and Nebulae.* Courtesy of John G. Wolbach Library, Harvard College Observatory.

material! Certainly inspired by the works of William Herschel and Pierre Simon Laplace, he later surmised boldly that the Andromeda "nebula" was a stellar system in formation. Clearly the Nebular Hypothesis had become a blinding mantra.

Roberts' passion for astrophotography was the fortunate cause of his meeting with a sister soul, the young American astronomer Dorothea Klumpke (1861–1942), who would play a unique legacy role in highlighting his pioneering astrophotographic work. In 1901, Roberts married Dorothea, who was 32 years his junior. Klumpke had obtained a doctorate degree in mathematics from La Sorbonne, Paris. When she met Roberts, she was Director of the Bureau of Measurements at the Paris Observatory and was in charge of the Carte du ciel, which was led by Paris Observatory Director, Admiral Mouchez. Klumpke and Roberts had met in 1896 while sailing to Norway with a group of astronomers, to observe a solar eclipse.

Once married, Dorothea left her position to work with Roberts. After Roberts' death in 1904, she completed his astrophotographic project by assembling the collection of photographic plates he had obtained at Crowborough. She published several papers as

Mrs. Roberts. In 1929, she released a comprehensive catalogue, *The Isaac Roberts Atlas of 52 Regions, a Guide to William Herschel's Fields of Nebulosity*, to help observers study extensive diffuse nebulosity.[32] It was an extraordinarily careful production; the photograph of the Messier 31 field is spectacular. In 1934, Dorothea received a crowning honor for her career. She was elected Chevalier de la Légion d'Honneur, and received the Cross of the Legion from the President of the Republic of France. Klumpke left several gifts of money to establish prizes in astronomy and mathematics.

There is one last consideration when regarding Roberts' work. His contributions are lesser known than those of Parsons. Perhaps Roberts' rush to interpret the "nebulae" as examples or confirmation of the Nebular Hypothesis made him a victim of whiggish historiography, where losers are forgotten and winners venerated.[33] By comparison, Parsons emphasized the power of his instruments and the quality of their images, and remained extremely cautious in his interpretation.

Unsettling "Grand Amateurs"

From William Herschel all the way to William Parsons, it can be seen that it was those individuals with mechanical genius and a knack for solid management who made and improved large reflectors and used them in creative ways. It seems remarkable that businessmen were able to move into the world of astronomy late in their life. Parsons made significant contributions not only in building the largest telescopes of the nineteenth century, but also in developing new and systematic approaches to astronomical observing and telescope operations. Also striking were the groundbreaking contributions of these individuals who transformed themselves into devoted astronomy grand amateurs and who audaciously pushed the technology of large reflectors and the use of astrophotography. They all shared a strong interest in "nebulae," which drove their technical innovation. Like Allan Chapman, Alan Hirshfeld has presented fine portraits of these inventors and eccentrics who have contributed in unusual and innovative ways to the progress of astronomy.[34] These outsiders unsettled the rather quiet professional astronomers.

While professional astronomers, working mostly in positional astronomy, pushed for the use of refractors that they found more tractable and precise, the grand amateur wanted larger instruments and pressed for increased aperture. Further, they improved telescope mount technology and added a piggyback viewing telescope for sidereal tracking during hour-long exposures. The innovative William Lassell was an inspirational figure for the American astronomer George Ellery Hale. Lassell built several reflectors of steadily increasing size and installed them on equatorial mounts, first using them at his own private observatory. The largest, a 48-inch reflector, he moved, with family and all, to Malta,

[32] D. Klumpke, *The Isaac Roberts Atlas of 52 Regions, a Guide to William Herschel's Fields of Nebulosity*. Diffuse nebulosity was detected in only four of the Herschel fields.

[33] S. Shapin and S. Schaffer, *Leviathan and the Air Pump, Hobbes, Boyle, and the Experimental Life*, Princeton: Princeton University Press, 1985, 2011 with a new introduction, pp. 8–9.

[34] A. Hirshfeld, *Starlight Detectives: How Astronomers, Inventors and Eccentrics Discovered the Modern Universe*, New York: Bellevue Literary Press, 2014.

seeking clearer skies, as Hale would later do by moving from Yerkes Observatory in Wisconsin to Southern California.

One telescope built by grand amateurs has had a particularly astonishing story. Andrew Common (1841–1903), a plumbing and heating engineer, built "the first of the line of successful large metal-film-on-glass modern reflectors," the Common's 36-inch reflector.[35] Common put this instrument to task for astrophotography at his home in Ealing, a suburb of London. He was quite successful, and he received the gold medal from the Royal Astronomical Society for his photographs of "nebulae." This was start of an extraordinary venture for his 36-inch reflector. As he proceeded to build a 5-ft reflector, Common sold his 36-inch to Edward Crossley (1841–1905), head of J. P. Crossley and Sons, a family-run textile mill in Halifax, Yorkshire.[36] In late 1885, Crossley transported his newly acquired telescope to Bermerside where he built a new spherical dome to shelter it better. The very busy Crossley employed a full-time astronomer and took part in the observations whenever possible. Disappointed by the poor English climate, Crossley decided to dispose of the reflector and dome.

After convoluted negotiations with the Regents of the University of California, finally completed in 1895, Crossley donated the telescope and dome to Lick Observatory, located on Mount Hamilton in the Sierra Mountains of California. He had to pay for a fair fraction of the moving expenses because the university was short of funds for a project it considered highly risky. By June 1896, the Crossley, as it came to be called, was installed at the Lick Observatory and made suitable for groundbreaking photographic investigations on "nebulae," and this opened an amazing new chapter.

As demonstrated by Parsons and Lassell, reflectors could be made big, and hence were more powerful for detecting and observing faint "nebulae." They would triumph in the twentieth century, while the great refractors became "extinct" soon after the arrival of the mammoth Yerkes Observatory 40-inch in 1895. With foresight, Isaac Roberts had predicted that great progress would be achieved in probing the sky by photography using big reflecting telescopes, which would bring astrophotography to full maturity. At the dawn of the twentieth century, the essential tools were in place to solve the thousand-year-old mystery of "nebulae."

B – PHOTOGRAPHING MYRIADS OF GALAXIES: THE GOLDEN AGE OF ASTROPHOTOGRAPHY

How Astrophotography Revealed the Distance and Nature of Galaxies

The increased use of photography, along with the improved sensitivity of photographic emulsions at the turn of the twentieth century, contributed to make astronomy a leading image science. As astronomers adapted to and adopted photography, telescopes were now

---

[35] R. P. S. Stone, The Crossley Reflector: A Centennial Review – I & II, *Sky and Telescope*, 1979, October issue, pp. 307–311; November issue, pp. 396–400.

[36] In 1933, the 5-ft was refurbished and set up as the 1.5-m Boyden–UFS reflector at Boyden Observatory in South Africa.

constructed to enable photography. Astrographs were designed to provide a large optically corrected field of view. During the first decades of the twentieth century, many of the promises alluded to by George Ritchey for "a broader science" were being realized in spectacular ways.

The Giant Steps of Astrophotography

A new generation of young astronomers took over Draper's, Common's and Roberts' pioneering work in nebular astrophotography in the nascent astronomy community of North America. James Keeler and Heber Doust Curtis (1872–1942) at the Lick Observatory on Mount Hamilton, central California employed the refurbished Crossley 36-inch reflector in most skillful ways.[37]

Having been shipped over and received at Lick, the 36-inch was in such a poor shape that it was declared by not just a few of the observatory staff as a "piece of junk," a "monstrosity." After a difficult transition period involving many repairs, the leading astronomer at the observatory, James Edward Keeler (1857–1900), successfully put it to work (Fig. 3.3).[38] However, it was American astronomer Charles Dillon Perrine (1867–1951) who completely redesigned and rebuilt the telescope, giving it its familiar look. Due to these diligent efforts, the telescope mount was rebuilt and the telescope transformed into a powerful astrograph that started operation in 1898.

The high mountain site also offered fine observing conditions. Thus equipped with an optically fast, very fine, glass mirror, Keeler and Curtis conducted a systematic photographic survey of the sky with a new wide-field photographic camera. Following the untimely death of Keeler at the age of 43 in 1900, Curtis vigorously carried out the Lick Observatory "nebulae" astrophotography program. The Crossley camera covered 0.9 degree at any one time, a field with an area almost ten times that of the apparent size of the Moon. With a survey covering only a few areas of the sky, the deep photographs revealed thousands of galaxies. Extrapolating from these counts, Curtis estimated in 1918 that one million or more spirals would be detectable with the Crossley, if the whole sky were photographed to the same depth.[39]

In 1918, Curtis published extensive papers accompanied by detailed descriptions of photographs of "nebulae" and star clusters made with the Crossley from 1898 until 1 February 1918 (Fig. 3.4). The lists contained 762 entries, of which 513 were classified as spirals. Curtis was quite clear about the nature of the latter objects: "It is my belief that all the many thousands of nebulae not definitely to be classed as diffuse or planetary are true spirals. . . .

[37] Built by Andrew A. Common in Great Britain in 1879, the Crossley was the first large reflector to be equipped with a concave, silver-coated glass mirror. Having been refurbished several times, the telescope is still in use today. For a fine historical review, see Remington P. S. Stone, The Crossley Reflector: A Centennial Review – I & II, *Sky and Telescope*, 1979, October issue, pp. 307–311; November issue, pp. 396–400.

[38] J. Keeler, The Crossley Reflector of the Lick Observatory, *Publications of the Astronomical Society of the Pacific*, 1900, Vol. 12, pp. 146–167.

[39] H. D. Curtis, The Number of the Spiral Nebulae, *Publications of the Astronomical Society of the Pacific*, 1918, Vol. 30, pp. 159–161.

Fig. 3.3 The Lick Observatory 36-inch Crossley reflector, still with the original structure from Crossley. From Keeler (1900). *The Astrophysical Journal.* © AAS. Reproduced with permission.

Fig. 3.4 Heber Curtis. Credit: Bentley Historical Library, University of Michigan.

Were the Great Nebula in Andromeda itself situated five hundred times as far away as at present, it would appear as a structureless oval about 0.'2 long, with very bright center, not to be distinguished from the thousands of very small, round or oval nebulae found wherever the spirals are found."[40] (Note that 0.'2 means 0.2 arcminute of angular size.)

Curtis' article provides interesting reading as it followed very much the tabular listing of objects of the reports of the Birr Castle works published by the Third and Fourth Earls of Rosse decades earlier. Similarly, Curtis employed short telegraphic descriptors to depict the various objects: for example NGC 7723 is "A nearly round, rather faint, open spiral of the Φ-type, 1'.5 long. Bright stellar nucleus. 6 s.n." (Note that Φ meant that it was classified as a barred spiral.) However, there was a stark difference with the earlier publications: Curtis' article was accompanied by a fine set of photographic plates illustrating the different forms of barred spirals. Despite Curtis' achievements, observing with the refurbished 36-inch remained challenging. American astronomer William Hammond Wright (1871–1959), director of the Lick Observatory from 1935 to 1942, wrote sarcastically "the only

[40] H. D. Curtis, Descriptions of 762 Nebulae and Clusters Photographed with the Crossley Reflector, *Publications of the Lick Observatory*, 1918, Vol. XIII, Part I, p. 12.

Fig. 3.5 George Ellery Hale. Credit: University of Chicago Photographic Archive, [apf6–00263], Special Collections Research Center, University of Chicago Library.

comfortable and safe way to work with the Crossley would be to flood the dome with water and observe from a boat."[41] But another team of American astronomers soon came on the scene with new revolutionary telescopes and cameras at a fine mountain site.

Trailblazer of Modern Astrophotography

During the first decade of the twentieth century, a gigantic effort in creating a completely new astronomical facility was taking place at Mount Wilson Observatory, Southern California. American astronomer George Ellery Hale (1868–1938), who understood the importance of good sites for astronomical work, was leading the colossal venture (Fig. 3.5). Originally at Yerkes Observatory in Wisconsin, Hale was a visionary astronomer with an unusual ability for promoting science and an incredible instinct for successful fundraising. He applied these talents to the creation of scientific institutions and the building of big

---

[41] Cited by R. P. S. Stone, The Crossley Reflector: A Centennial Review – I & II, *Sky and Telescope*, 1979, November issue, p. 400.

telescopes.[42] Hale also understood well that the size of telescope was not everything. Good instruments and a creative staff were needed as well. Having raised ample amounts of money, he swiftly used these funds, hiring researchers at the forefront and constructing superb astronomical facilities equipped with the best cameras and spectrographs.

Among the newly hired staff was George Willis Ritchey (1864–1945). Ritchey was a first-class engineer and inventor who saw the enormous advantages of the arrival of large blank disks of glass that could be configured and polished to an extremely smooth and precise optical surface. Following the successful techniques developed by French astronomer Léon Foucault (1819–1868) at Marseille Observatory, he knew that the glassy surface could also be coated with a thin layer of silver for optimal reflection – as had been done with the Crossley. Because of their superiority in several physical and technical aspects, glass mirrors supplanted metal ones (and glass lenses) as soon as they could be cast at meter size and larger. In addition, Ritchey redesigned mechanical structures for telescope mounts and drives appropriate for large reflectors. These innovations resulted in spectacular improvements in the optical quality of telescopes and in the precision of their mechanical movement.

Ritchey had understood the requirements and potential of photography. He spearheaded the new technology of photography in unique ways. While at Yerkes Observatory, the institution that then hosted the largest refractor in the world, he designed and constructed a fast (f/4) 24-inch reflector in 1901 (Fig. 3.6). He conducted illustrative demonstrations highlighting his design. Until then, only grand amateurs favored reflectors, if we exclude Léon Foucault in France. "Ritchey took a spectacular series of photographs of nebulae with it [the 24-inch], demonstrating to skeptical astronomers just how useful reflecting telescopes could be."[43] Having moved to Mount Wilson, Ritchey became lead designer for the new telescopes being built at the new California site. He designed two scientific monuments optimized for deep photographic work: the 60-inch and in part the 100-inch reflecting telescopes of the Mount Wilson Observatory (Figs. 3.7a and 3.7b). Francis Gladheim Pease (1881–1938), who had been Ritchey's assistant, replaced him as project leader on the 100-inch, for reasons we will see later. The 60-inch was the first great modern reflector. Many considered it an engineering marvel.

Together, Hale and Ritchey made a powerful scientific tandem that lasted for a few decades. They left a huge and rich legacy. The fruits of the Hale–Ritchey alliance were an extraordinary set of new astronomical facilities capable of exploiting the exquisite conditions of the mountain sites of southern California with the added power of astrophotography. Due to Hale's excellent management and diligent work, the 60-inch reflector was in operation before the end of 1908. Combining passion with a fanatic sense of precision, Ritchey came with a background that made him an outstanding astrophotographer. He used the 60-inch to obtain deep photographs of "nebulae," spirals in particular. Many of Ritchey's photographs were obtained through very long exposures: to reach as deeply

[42] J. R. Goodstein, *Millikan's School*, New York: W. W. Norton and Company, 1991, pp. 64–87.

[43] D. E. Osterbrock, *Yerkes Observatory 1892–1950: The Birth, Near Death, and Resurrection of a Scientific Research Institution*, Chicago: University of Chicago Press, 1997, pp. 36–37.

Fig. 3.6 George Ritchey's 24-inch reflector (1933). Credit: University of Chicago Photographic Archive, [apf6–01377], Special Collections Research Center, University of Chicago Library.

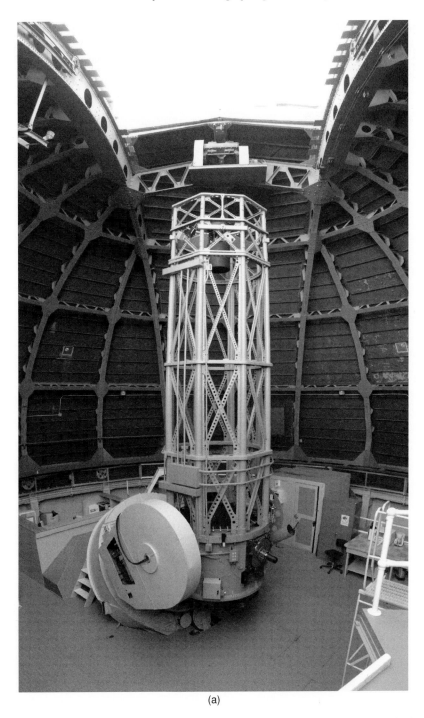

(a)

Fig. 3.7 (a) Century-old modernized 60-inch reflector at Mount Wilson Observatory. Credit: Creative Commons Attribution/Heven729.

(b)

Fig. 3.7 (b) The Mount Wilson 100-inch reflector as seen today. Credit: Creative Commons Attribution/Ken Spencer.

as possible, exposures on the same plate were spread over several nights, in one case total-ing 60 hours. In their 1935 classic work, *Lunettes et télescopes*, French astronomers André Danjon and André Couder wrote that the photographs obtained by Ritchey with the Mount Wilson Observatory 60-inch reflector 25 years earlier were still the best available.[44]

Astrophotographer Guru

To highlight the excellence of the 60-inch optical design and its mechanical accuracy, Ritchey repeated what he had done with the Yerkes 40-inch refractor and his 24-inch reflec-tor, producing spectacular photographs of the Moon to demonstrate the quality and power of the new large reflector.[45] The images also confirmed the importance of locating such instruments on dry, high mountain sites such as those of southern California. Indeed, these high mountain sites provided not only extensive cloudless periods, but also benefited from the stabilizing effects of the cool deep waters of the Pacific Ocean. Long periods of uninter-rupted air stability led to finer images, better than anything achievable at low-altitude sites. Gone was the "boiling" atmosphere that affected the Birr Castle observers.

Ritchey's photographs surpassed, in their depth and definition, anything that had been done before. This allowed him to tackle the Gordian knot of the distances to "nebulae." He started searching for variable objects in them. In photographing the "nebulae," he found and followed relatively rare types of explosive stars, novae. Several of these had been observed in the Milky Way over the previous decades and their light curves were well established. Novae are not regular pulsating stars. Their brightening is triggered by the "dumping" of material from a red-giant companion onto a white dwarf. A nova corresponds to the flash of light, ignited by catastrophic nuclear detonation at the surface of the receiving star. The stellar blast causes a rapid brightening of several magnitudes, a phase that lasts for sev-eral days to a few weeks. Hence, novae could be registered on photographic plates taken only weeks apart. Although he could then not have known the nova mechanism, Ritchey declared correctly that novae were as numerous in Messier 31 as they were in the Milky Way, about a dozen events a year. The objects appeared very faint, he explained, because of their much greater distances. He followed a nova in Messier 31 over several nights. He also found a nova in Messier 81, two in Messier 101 and one in NGC 2403.[46] Assuming they were analogues of novae in our own Milky Way and scaling for their difference in brightness, he derived distances to the host systems to be millions of light-years! Ritchey became convinced of the extragalactic nature of most "nebulae."

Donald Osterbrock summarized Ritchey's main abilities as follows: "superb tech-nical skill and instrumentation, striking photographs, and very few hard scientific

---

[44] A. Danjon and A. Couder, *Lunettes et télescopes*, Paris: Librairie scientifique et technique Albert Blanchard, 1999, p. 694 (original work published in 1935).

[45] A collection of George Ritchey's photographs (including glass plates) were auctioned by Bonhams for about half a million US dollars in 2014.

[46] "Novae" were not all of the same nature: those in Messier 81, 101 and NGC 2403 were actually supernovae, which are much more powerful and brighter exploding stars than ordinary novae, as demonstrated by Rudolph Minkowski and Fritz Zwicky (1941).

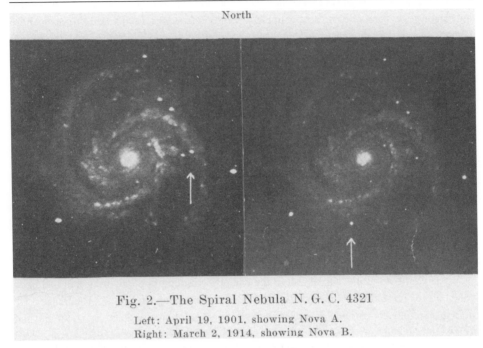

North

Fig. 2.—The Spiral Nebula N. G. C. 4321

Left: April 19, 1901, showing Nova A.
Right: March 2, 1914, showing Nova B.

Fig. 3.8 **Transformational Image: Photographs of "Novae" in Spirals.** Supernovae in NGC 4321 that Curtis thought were novae, significantly less luminous objects. From Curtis (1917), *Lick Observatory Bulletins*.

results."[47] Indeed, the procrastinating Ritchey did not proceed fast enough with publishing his results, probably not having realized the significance of his novae finding. Curtis at Lick Observatory did otherwise. Rushing to the 36-inch plate archive of spirals, Curtis found several more novae. He quickly prepared and published two excellent papers before Ritchey managed to do so (Fig. 3.8).[48] From these "novae," he argued spirals to be 20 million light-years away and 60,000 light-years in diameter.

Images and the Astronomical Time Domain

By the end of the first decade of the 1900s, astronomers had available a new generation of superb telescopes located at outstanding sites of mountainous California, Mount Wilson in particular. The conditions of air stability that prevailed at these high sites helped to finally establish the superiority of photography over visual observing and hand drawing. By

---

[47] D. E. Osterbrock, *Pauper & Prince: Ritchey, Hale & Big American Telescopes*, Tucson: University of Arizona Press, 1993, p. 52.
[48] H. D. Curtis, Three Novae in Spiral Nebulae, *Lick Observatory Bulletin*, 1917, Vol. 9, Number 300, pp. 108–110.

developing telescope systems for long exposures, astrophotographers finally demonstrated that photographs were much more reliable recorders of astronomical features, especially of flimsy objects such as "nebulae." The reliability of the photograph quickly surpassed that of drawings, even when supplemented by extensive notes. Photography was especially useful and faithful at registering elusive features that the eye could never have registered nor sketched reliably. The amount of darkening on the emulsion could be related quantitatively to the brightness of objects. Photographs did not need much of a commentary; just a short caption and a description of the technical set-up sufficed.

In the light of these new achievements, the excruciating work of Birr Castle observers appeared now to be quite obsolete. At the same time, it was realized how meritorious their findings had been. This impression was reinforced by the confirmation that what the Earl of Rosse's team had recorded turned out to be relatively faithful images of "nebulae." Commenting on Parson's drawings of Messier 31 (Andromeda Galaxy) executed with the Leviathan, Isaac Roberts wrote: "These make one acquainted with the difficulties encountered by the observers in their efforts to make intelligible, by descriptive matter and by drawings, an object having a structure so complex that no eye and hand, however well trained, could possibly delineate it."[49]

It was not only a matter of a more faithful recording of features. Photographs made it easier to compare a new observation of a given object with a previous observation. Time-domain observing was put on a solid footing. With novae, Ritchey and Curtis had shown the power of astrophotography in discovering and monitoring time-variable phenomena.[50] One just needed to compare images taken a few nights, weeks or months apart. Furthermore, the relatively large field of view afforded by photographic plates provided a number of stars that could be used as coordinate references and calibrators of brightness. Calibration ensured that quantitative physical information was derived from photographs. The brightness of new stars could be established from other reference stars outside the field; the latter became the "standard stars," specific objects photometrically calibrated through a rigorous procedure. This was the origin and driver of the massive photometric and spectroscopic surveys of Harvard College Observatory.

An Unruly Perfectionist

Unfortunately, the eccentric and individualistic Ritchey ran into trouble with his colleagues. Ritchey was an absolute perfectionist. "An artist rather than a 'bench' research scientist, Ritchey took infinite pains to achieve perfection. If he thought a certain piece of equipment would help perfect the final product, he made request after request until he obtained the item."[51] Furthermore, in his mind "the goal of a telescope was to produce pretty pictures

---

[49] I. Roberts, Photograph of the Spiral Nebula M33 Trianguli, *Monthly Notices of the Royal Astronomical Society*, 1895, Vol. 56, p. 70.

[50] G. Ritchey, Novae in Spiral Nebulae, *Publications of the Astronomical Society of the Pacific*, 1917, Vol. 29, pp. 210–212.

[51] A. R. Sandage, *Centennial History of the Carnegie Institution, Volume 1: The Mount Wilson Observatory*, Cambridge: Cambridge University Press, 2004, p. 169.

of the highest technical quality, regardless of any other use the images might have for a sci-entific project."[52] Therefore, it was no surprise that the arrogant Ritchey became his own worst enemy, the victim of his own sense of importance. American astronomer and astron-omy historian Donald Osterbrock described the tandem of Ritchey and Hale as "pauper and prince."[53] Around 1912, Ritchey fell from grace.

Ritchey's egotist personality, his grandiose visions and failures at delivery on promises were factors that had made relations with his Mount Wilson colleagues turn sour. He had unwisely bypassed Hale in approaching donors, and had misrepresented his role and author-ity too many times. In reaction, Hale and Walter S. Adams treated him quite unfairly by for-bidding him access to the 100-inch, by blocking his nominations to prestigious prizes and by censoring publications that highlighted Ritchey's achievements. "Ritchey was unsparing of himself and of those who worked with him. He believed that he alone knew the secrets of making large telescopes; he was contemptuous of rivals like John A. Brashear and Francis Pease. Ritchey was tremendously self-centered, and in the end his failure to accommo-date himself to the vision of this director cost him his job at Mount Wilson."[54] In 1919, following years of insubordination and disagreements with Hale and with the succeeding director Walter S. Adam, as well as with other staff, he was dismissed from Mount Wilson Observatory.

Ritchey's Wide-Field Imaging Legacy

After leaving Mount Wilson Observatory, Ritchey went to France where he worked with French optical designer Henri Chrétien (1879–1956). Both understood the power of wide-field photography, as they pushed for bigger fields of view. Being the perfectionist he was, Ritchey worked to correct optical aberrations. Mirror systems with parabolic mirrors led to coma, a simple optical aberration that produces elongated images of point sources away from the optical axis. Working together, Ritchey and Chrétien developed a radically new concept of telescope optics, with new curving surfaces that eliminated coma (Fig. 3.9; com-pare with Fig. 3.6). They designed primary and secondary mirrors with hyperbolic shapes instead of parabolic ones. A result was that the shape of images remained perfect over a large field view in the focal plane, empowering even more astrophotography in catching a greater patch of the sky in each exposure. Ritchey's later work was devoted to designing future very large telescopes with primary mirrors up to 8 m in diameter. Chrétien is also known as the inventor of cinemascope and was awarded a Hollywood "Oscar" from the American Academy of Motion Pictures Arts and Science in 1954.[55]

---

[52] A. R. Sandage, *Centennial History of the Carnegie Institution, Volume 1: The Mount Wilson Observatory*, Cambridge: Cam-bridge University Press, 2004, p. 169.

[53] D. E. Osterbrock, *Pauper & Prince: Ritchey, Hale & Big American Telescopes*, Tucson: University of Arizona Press, 1993.

[54] D. E. Osterbrock, *Pauper & Prince: Ritchey, Hale & Big American Telescopes*, Tucson: University of Arizona Press, 1993, pp. 284–285.

[55] Henri Chrétien received the Academy Award of Merit along with Earl Sponable, Sol Halperin, Lorin Grignon, Herbert Gragg, and Carlton W. Faulkner: "For creating, developing and engineering the equipment, processes and techniques known as [20th Century Fox's] CinemaScope." The award was also shared with Fred Waller, "For designing and developing the multiple photographic and projection systems which culminated in Cinerama."

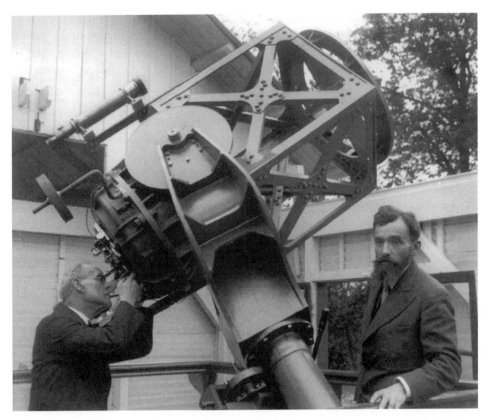

Fig. 3.9 George Ritchey and Henri Chrétien with a 20-inch telescope of their design in 1927. © Françoise Le Guet-Tully, Fonds Henri Chrétien, with permission.

Throughout the twentieth century, larger photographic fields and better images became a never-ending challenge and goal for telescope and astronomical camera designers and optical engineers. Most telescopes nowadays, including the Hubble Space Telescope, use the Ritchey–Chrétien design. Notwithstanding the unhappy ending at Mount Wilson, it is impossible to overstate the technical mastery and the innovative concepts of the great engineer and telescope designer that Ritchey was. "Ritchey was an outstanding telescope maker and the archetype of all telescope makers."[56] In several ways, his obsession for beautiful, large-scale and finely resolved telescopic images paid off.

In the 1920s, the 60-inch and 100-inch telescopes at the Mount Wilson Observatory became the fundamental tools to confront the foremost problems of the nature of galaxies and of the structure of the universe. The mapping of our own Milky Way received a boost. The plates obtained with the 60-inch helped to establish the numbers of stars to fainter

---

[56] D. E. Osterbrock, *Pauper & Prince: Ritchey, Hale & Big American Telescopes*, Tucson: University of Arizona Press, 1993, p. 283.

magnitudes and to reconstruct the distribution of stars in our galaxy. William Herschel's dream was achieved. "It was the culmination of the Mount Wilson program to provide basic data on faint stars for the Kapteyn program."[57]

The new large reflectors and cameras also represented technological and operational breakthroughs that paved the way a decade later for the groundbreaking work of Edwin Hubble, Milton Humason, Fritz Zwicky and Walter Baade. Using the 60-inch and the 100-inch, Edwin Hubble was able to confirm the extragalactic nature of the majority of the "nebulae.". In Chapter 6, I show how the great telescopes enabled the ultimate breakthrough in the exploration of galaxies.

As companies, especially Eastman Kodak, developed new emulsions for the specific needs of astronomy, astrophotography underwent several major improvements throughout the twentieth century, which provided both finer resolution and increased sensitivity of the photographic plate. Creative processing techniques also helped greatly. British–Australian astronomer David Malin, a chemist and microscopist in his early career, was among the few who pushed forward with innovative techniques: the stacking of exposures from several plates to increase the signal-to-noise ratio; unsharp masking to reveal faint structures barely above the level of the sky background; and the production of three-colour wide field images.[58] In the late 1980s, electronic digital detectors largely superseded silver-based photographic emulsions. The main technology employed charge-coupled devices (CCDs), which were several times more sensitive. In addition, their response is linear (the signal is proportional to the amount of light hitting the detector) and they can now be assembled in mosaics larger than the largest photographic plates. Nevertheless, several of the original photographic processing techniques were successfully adapted, improved and optimized for use with electronic images.

Scientific Drawing in the Modern Age

One may well then ask: "With the advent and universal use of photography, has astronomical drawing become obsolete?" Certainly not, as various forms of sketching have remained useful for scientific illustration, astronomy being no exception. Swiss astronomer Fritz Zwicky (1898–1974) gave fine examples of how drawing could be used effectively to complement photography. Working at the California Institute of Technology, Zwicky remains well known today for surveys of clusters and groups of galaxies and for several other brilliant ideas. A Swiss astronomer working at the California Institute of Technology and a frequent user of the Mount Wilson Observatory, and later of the Palomar Observatory, Zwicky was a very creative and innovative astrophysicist. He made important contributions to twentieth-century astrophysics.[59] He is credited with several revolutionary concepts, such

---

[57] A. R. Sandage, *Centennial History of the Carnegie Institution, Volume 1: The Mount Wilson Observatory*, Cambridge: Cambridge University Press, 2004, p. 225.

[58] D. Malin and P. Murdin, *Colors of the Stars*, Cambridge: Cambridge University Press, 1984.

[59] A. Stöckli and R. Müller, *Fritz Zwicky: An Extraordinary Astrophysicist*, Cambridge: Cambridge Scientific Publishers, 2011.

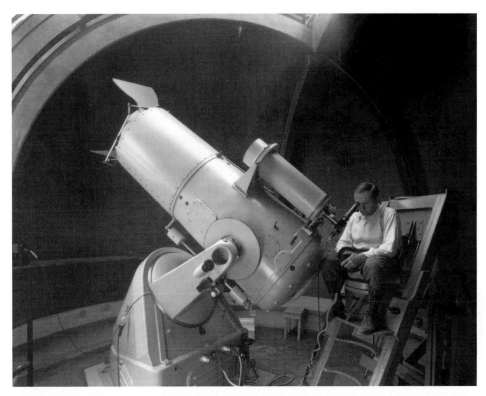

Fig. 3.10 Fritz Zwicky at the Palomar 18-inch Schmidt Telescope in 1936. Credit: Courtesy of the Archives, California Institute of Technology.

as neutron stars, gravitational lenses, interacting galaxies and for the first unassailable evidence of the prevalence of dark matter in the universe (Chapter 8).

Zwicky had been using the light-tight Palomar Mountain 18-inch and 48-inch Schmidt telescopes for photographic surveys of galaxies (Fig. 3.10). The name of the telescope optics design recognizes German optician Bernhard W. Schmidt (1879–1935), who invented the concept. The Schmidt design combines a Cassegrain reflector configuration and a large corrector plate, which results in a compact telescope that gives a super-wide field of view. Schmidt telescopes, although of small aperture, provide significantly larger fields of view than even the Ritchey–Chrétien design. Fields of view several times the angular size of the Moon can be imaged in a single shot. Schmidt telescopes were most suitable for Zwicky's work because galaxy groupings were of larger angular sizes than the fields of view of large telescopes, such as the Mount Wilson 60-inch and 100-inch.

Zwicky had found a "surprisingly large number of rather widely separate galaxies which appeared connected by luminous intergalactic formations."[60] These structures were very

[60] F. Zwicky, Multiple Galaxies, *Ergebnisse der exakten Naturwissenschaften*, 1956, Vol. 29, pp. 344–385.

faint, showing barely above the brightness level of the sky background. A very tight light bucket was needed to detect them. In terms of light baffling, the powerful Mount Wilson 100-inch was like an open door; it could not record the faint intergalactic features until all sources of stray light were eliminated. Hence, it was very challenging to bring out faint features at the limit of the photographic emulsion. Moreover, it was extremely difficult to reproduce the elusive features on a positive print and even more challenging to get them to come out in the printed journals. These limitations led Zwicky to draw the features by hand. He used simple sketches to highlight "multiple galaxies" and the faint filamentary structures that connected some of the galaxies, structures that he interpreted correctly as the result of inelastic collisions.[61]

Zwicky also employed a sequence of drawings to illustrate how interacting galaxies could produce bridges and tails of stars as material was torn apart from the parent galaxies. In a seminal drawing, he described the transfer of momentum and the formation of tidal tails as two galaxies pass by each other and interact (Fig. 3.11; see also Fig. 6.8). These examples from Zwicky's work represent a fine example of the use of sketching to illustrate a new phenomenon, based on the detection of features then at the very limit of instrument capabilities. The sketches were somewhat a reversal from the Worthington problem: While Worthington used photography to show that his earlier drawings had idealized the shapes of splashing drops, Zwicky employed old-style drawing to highlight real features barely visible on the original photographic plates.

It is compelling that for his 1953 article in *Physics Today*, Zwicky chose to show a sketch of the double galaxy Messier 51.[62] For the Swiss–American Carnegie Observatories astronomer François Schweizer, Zwicky's "sketch shows faint details that became visible to most folks only later when the new Kodak IIIa-J plates showed significantly fainter details than the old 103a-0 or IIa-O plates did."[63]

Today's astronomers, professional and amateur, have pushed the art and science of astrophotography to new heights.[64] Researchers using images obtained with the Hubble Space Telescope and modern ground-based telescopes are now producing superb colour images by combining images obtained in filters of different wavelength passes. These images are used to derive important scientific measurements. The images are also a powerful means to convey the beauty of astronomical objects and to share the excitement of discovery with a larger public. Additionally, "when processed correctly, an attractive and evocative picture brings out the scientific content within."[65] Amateur astronomers have also caught up spectacularly with their efficient equipment and advanced image processing with computers. They now produce images of a quality that were not even dreamt of by professional astronomers of a few decades ago.[66]

[61] F. Zwicky, Multiple Galaxies, *Ergebnisse der exakten Naturwissenschaften*, 1956, Vol. 29, pp. 366–370.
[62] F. Zwicky, Luminous and Dark Formations of Intergalactic Matter, *Physics Today*, 1953, Vol. 6, pp. 7–11.
[63] Private e-mail note to the author (Sept. 12, 2014).
[64] T. A. Rector, et al., Image-Processing Techniques for the Creation of Presentation-Quality Astronomical Images, *Astronomical Journal*, 2007, Vol. 133, pp. 598–611.
[65] R. Villard and Z. Levay, Creating Hubble's Technicolor Universe, *Sky & Telescope*, 2002, September issue, pp. 28–34.
[66] R. Gendler, Forays into Astronomical Imaging: One Person's Experience and Perspective, *Astronomy Beat*, Astronomical Society of the Pacific, 2011, No. 79, August 30, pp. 1–6.

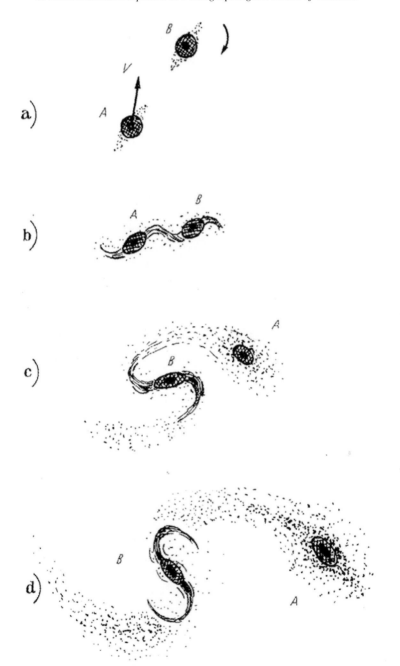

Fig. 3.11 Drawing sequence by Zwicky highlighting the dynamical phases of an interacting pair of galaxies and the formation of tidal tails. From Zwicky (1956), *Ergebnisse der exakten Naturwissenschaften*.

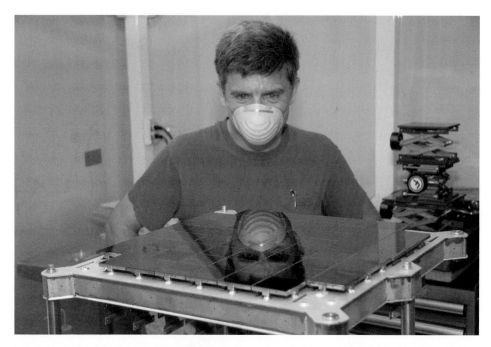

Fig. 3.12 Each of the four PanSTARRS cameras is equipped with a mosaic of 64 x 64 CCDs, spreading over an area of about 40 cm and providing a total of 1.4 gigapixels. Credit: Institute for Astronomy University of Hawai'i.

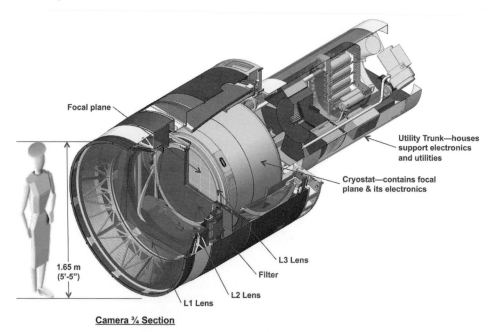

Fig. 3.13 Schematic of the LSST camera. Credit: LSST Corp./National Science Foundation.

A new phase of systematic photography of the sky is beginning. The 1.8-m Panoramic Survey Telescope and Rapid Response System (PanSTARRS) and the 8-m Large Synoptic Survey Telescope (LSST) will transform our view of the universe by fully exploiting the time domain (Fig. 3.12). Located in the northern hemisphere on Haleakala, Maui Hawai'i, the PanSTARRS is a 1.8-m wide-field telescope, which can observe the entire available sky several times each month. Built by the US Department of Energy for the southern hemisphere (site of Cerro Pachón, Chile), the 2.8-ton camera of the 8-m LSST blows the mind. Providing a 3.5-degree field of view with 189 sixteen megapixel CCDs, a single image has a total of 32 gigapixels (Fig. 3.13). The LSST science plan is to image most of the sky visible from the southern hemisphere through several filters a few times a month.

4

Portraying "Nebulae" for the Mind

Often the most effective way to describe, explore and summarize a set of numbers – even a very large set – is to look at pictures of these numbers.

Edward R. Tufte[1]

Photographs and naked-eye drawings of the Milky Way, however, must picture somewhat different portions of the stellar world.

Cornelis Easton[2]

When one becomes more familiar with the ordering of the hundred-odd chemical substances within the table, the symmetries seem so obvious, the sequences so natural, that most people find hard to imagine a time when this object did not exist...

Michael D. Gordin[3]

What is the Role of Abstract, or Representational, Images in Unveiling the Underlying Physics of "Nebulae"?

Ebenezer Porter Mason (1819–1840) was a young American astronomer who died from tuberculosis when only 21 years old. He is little known. John Herschel admired Mason's accurate and methodical work on "nebulae." He wrote, "Mr. Mason, a young and ardent astronomer, a native of the United States of America, whose premature death is the more to be regretted, as he was (as far as I am aware) the only other recent observer who has given himself, with the assiduity which the subject requires, the exact delineations of nebulae, and whose figures I find at all satisfactory."[4]

Mason was working with college friends to learn more about "nebulae." To conduct their pioneering work, Mason and Yale College friends built a 30-cm reflector, at the time the largest telescope in the Americas. Mason's goal was to advance the study of "nebulae"

---

[1] E. T. Tufte, *The Visual Display of Quantitative Information*, Cheshire: Graphic Press, 1983, p. 9.

[2] C. Easton, A Photographic Chart of the Milky Way and the Spiral Theory of the Galactic System, *The Astrophysical Journal*, 1913, Vol. 37, p. 105.

[3] M. D. Gordin, A *Well-Ordered Thing, Dmitrii Mendeleev and the Shadow of the Periodic Table*, New York: Basic Books, 2004, p. xvii.

[4] J. Herschel, *Results of Astronomical Observations Made During the Years 1834, 5, 6, 7, 8 at the Cape of Good Hope*, London: Smith, Elder & Co., p. 1847.

beyond just description and cataloguing. He was involved in geodesic work, participating in the field survey of the Maine–Canadian border. As we will see later, his knowledge of topography may have influenced his approach to astronomical work, and nebular observing in particular. He applied the technique of contour lines to his study of "nebulae." It was a first step in representational imaging to study "nebulae" which would become a most powerful tool of twentieth-century astrophysics.

Representing the Immense with Synoptic Imaging

The celestial vault is immense and unfathomable. As observing tools, telescopes, photographic plates and camera systems improved and became more powerful, the number of astronomical objects recorded increased phenomenally. Let us recall the Carte du ciel and Astrographic Catalogue project. Already in 1918, Heber Curtis had completed a rapid photographic survey of the sky with the 36-inch Crossley telescope at the Lick Observatory. He had inferred the number of spiral galaxies in the observable volume of the Crossley to be at least one million.[5] Beyond the sheer number, astronomical photography revealed a staggering amount of details and features. Astronomers also found many new forms of sidereal objects: asteroids in huge numbers, patches of darkness or "empty regions," groups and clusters of objects, nebulosities of many kinds and shapes and galaxies in ever increasing numbers.

It is no surprise that researchers felt the need to integrate the vast amount of information in a more synthetic form, which could be visualized differently. Synoptic charts and maps came to the rescue; they were effective means to summarize, average and distil large amounts of information. As implied by Edward Tufte's quote, new images were created, but they were images of the mind and for the mind, not of a natural phenomenon.[6]

Synoptic charts and maps are called non-homomorphic representations, as opposed to direct, homomorphic, images that the eye sees.[7] Geographical maps are the simplest and most familiar expressions of non-homomorphic representations; they may show contours of equal heights to emphasize and quantify the geographical relief of the Earth's continents or seafloors. Meteorologists make extensive use of synoptic maps of temperature and pressure distributions across land and sea for weather forecasts. There are also maps of hours of sunshine, height of snow or rainfall, foliage coverage and much else that integrate data over long periods of time.

Non-homomorphic representations are visual forms, visual "languages," that integrate or summarize large quantities of data, which can also be handled in tabular form. Images of transformed data are created to impress on the mind and to convey quantified information in a single synthetic view. Dmitrii Mendeleev's periodic table of chemical elements may be considered as one of the finest and most powerful non-homomorphic scientific

---

[5] H. D. Curtis, The Number of the Spiral Nebulae, *Publications of the Astronomical Society of the Pacific*, 1918, Vol. 30, 159–161.
[6] E. T. Tufte, op. cit., 1983.
[7] P. Galison, *Image & Logic, A Material Culture of Microphysics*, Chicago: University of Chicago Press, 1997, p. 19.

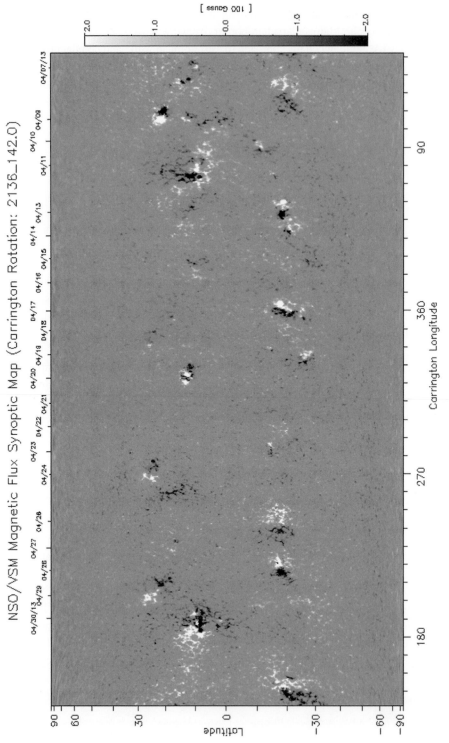

Fig. 4.1 Synoptic chart of the solar magnetic field assembled from individual magnetograms covering one full solar rotation in April 2013. Light shading shows the positive magnetic regions, and dark shading the negative regions. Credit: National Solar Observatory Integrated Synoptic Program.

representations of modern chemistry.[8] Maps of compiled data are now extensively used and produced; they serve many practical purposes. In the social or economic sciences, non-homomorphic representations help visualize vast sums of data: demographic indicators of population, natural resources, or the spread of endemic diseases, etc.

Because of their efficient summarizing power, non-homomorphic representations have been used extensively in the physical and natural sciences to synthesize or illustrate complex sets of data. They average several features in order for the intellect to process large quantities of information and to comprehend reality at a higher level. The photograph might raise an unneeded barrier for the neophyte. Geological maps and geological sections are highly complex, abstract and formalized kinds of representations.[9] Photographs of the terrain would not necessarily or easily reveal the richness and complexity of the geomorphological or stratigraphic landscape. The maps and charts are used to highlight features and guide the researcher or the student through the complex natural scenery.[10,11] Remarkable insights into geohistory have emerged by using higher-level representations.

Mapping the Heavens by Counting

In astronomy, the role of synoptic images has been to bring together or to average quantities of data or to provide a unifying picture of the systems or phenomena observed. For example, synoptic maps of the solar surface magnetic fields have been very useful for visualizing the large-scale magnetic properties of the Sun, for revealing the reversal of the overall magnetic field polarity every 11 years, and for making sense of the full solar activity 22-year cycle (Fig. 4.1). On a larger scale, objects external to the Milky Way appear much more numerous at high galactic latitudes. Edwin Hubble produced a "zone of avoidance" map to illustrate the obscuring material in the main plane of our Milky Way (Fig. 4.2; see Plate 6.1).[12] This obscuration limits our viewing ability in several directions as dust clouds hide a significant part of the universe at distances greater than a few hundred parsecs.

An early example of a synoptic representation in astronomy was the 1785 outline of the Milky Way star system by William Herschel (Fig. 4.3). His schematic model illustrated his ambitious approach of "constructing the heavens." Herschel observed countless stars, at whatever the direction he pointed his telescope. Being an astute observer, he noticed differences in the distribution of stars across the sky, finding variations subtler than the naked eye could perceive. Consequently, Herschel set himself the staggering task of reproducing the full three-dimensional distribution of the stars as he could see them with the aid of his telescopes. He decided to count stars in different directions, to make a census of all the stars in each given beam.

---

[8] M. D. Gordin, A *Well-Ordered Thing, Dmitrii Mendeleev and the Shadow of the Periodic Table*, New York: Basic Books, 2004.

[9] M. J. S. Rudwick, The Emergence of a Visual Language for Geological Science 1740–1840, *History of Science*, 1976, Vol. XIV, p. 159.

[10] T. Sharpe, The Birth of the Geological Map, *Science*, 2015, Vol. 347, pp. 230–232.

[11] M. J. S. Rudwick, *Earth's Deep History: How It Was Discovered and Why It Matters*, Chicago: University of Chicago Press, 2014, p. 140–142.

[12] E. P. Hubble, The Distribution of Extra-Galactic Nebulae, *The Astrophysical Journal*, 1934, Vol. 79, pp. 8–76.

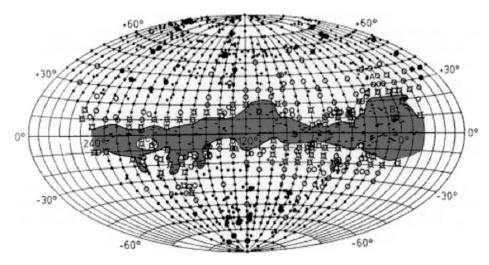

Fig. 4.2 Dust in the plane of the Milky Way absorbs stellar light and blocks our view to distant portions of the universe. Very few galaxies can be seen in that part of the sky, hence the appellation of "zone of avoidance." The zone of avoidance is sketched against the distribution of galaxies across the sky. From Hubble (1934), *The Astrophysical Journal.* © AAS. Reproduced with permission.

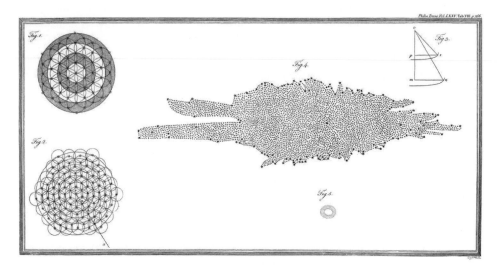

Fig. 4.3 Schematic of the "sidereal system" derived by William Herschel from his stellar counts. From Herschel (1785), *Philosophical Transactions of the Royal Society of London.*

To frame his endeavour soundly, Herschel made two simplifying assumptions: first, that his instruments allowed him to see into the farthest limits of the sidereal system; and secondly, that the stars were distributed regularly throughout space. He then defined approximately 300 sections having the same solid angle, spread across the sky that was visible from his observing location. Then he counted the stars in each of these conic sections. He counted and counted, using the brightness of the stars as an indicator of their distance, assuming that the fainter ones were further away. Herschel's method was primitive and his assumptions were incorrect. But overall his statistical sampling procedure was valid. The Kapteyn program that I discussed in Chapter 3 was a rebirth of Herschel's dream.

Herschel's counting method was adopted and improved by other astronomers during the nineteenth century, including his son John Herschel. It deeply influenced the work of Russian astronomer Friedrich Georg Wilhelm Struve (1793–1864), director of the Pulkovo Observatory in St. Petersburg.[13] Applying a similar counting method, Struve demonstrated that the brighter stars tended to conglomerate in the plane of the Milky Way. In 1847, Struve even suggested from his counts the existence of an attenuation of stellar light by one magnitude per c. 3,000 light-years of the line of sight. Although he did not explain the extinction mechanism, Struve had discovered an important component of the matter filling the space between the stars: fine dust particles distributed between the stars that absorb or diminish the intensity of stellar light as it travels through sidereal space, causing the so-called "zone of avoidance." This dimming effect was ignored for the next 70 years.[14]

In the 1910s, Heber Curtis used Herschel's statistical counting approach to estimate the number of spiral galaxies observable with the 36-inch Crossley reflector. Always doubtful of the reality of the large recession velocities of galaxies, Edwin Hubble conducted a systematic count of galaxies as a function of apparent brightness during the 1930s. He tried to use the number of galaxies counted to discriminate between models that explained galaxy spectral redshift as being due to the expansion of the universe, and models proposing instead some unknown phenomenon such as light "fatigue."

Confronting Bias

Naturally there is a subjective aspect to synoptic representation. For example, decisions have to be made about what features are important or what quantities need to be illustrated. As large set of data points are averaged, and a consistent method for smoothing the data has to be adopted and described. Furthermore, all scientific measurements are subject to biases. The designer needs to state the appropriate warning and limitations. For example, the use of a particular telescope, instrument and detector biases the astronomical detection in ways that depend on the observing wavelength, and hence the filter used. Biases and systematic effects are introduced by optical effects (e.g. diffraction or limits of the optics), quantum physics effects (threshold of detection or sensitivity as a function of wavelength), the duration of exposure (fainter objects remain undetected), and varying atmospheric conditions. As noted by James Keeler, the difference in the response of the human eye compared to that of the photographic emulsion explained some of the differences between objects recorded

---

[13] J. North, *Cosmos: An Illustrated History of Astronomy and Cosmology*, Chicago: University of Chicago Press, 2007, p. 448–449.

[14] Modern estimates of interstellar extinction in the Milky Way are 0.7–1.0 magnitudes per kiloparsec, or a factor of 2 to 2.5 of attenuation per 3,200 light-years.

by drawing and by photography.[15] The use of different photographic emulsions has been driven both intentionally and by practical reasons. As we will see in Chapters 9 and 10, such choices affect the process of galaxy classification and influence interpretation.[16]

Another biasing effect is brought about by sampling. For example, as we observe further and further away in the universe, intrinsically faint objects drop out. A sample of distant objects gets skewed towards the more luminous objects, as intrinsically brighter objects can be observed further away. This creates a false trend of increasing average luminosity as a function of distance. This is called the Malmquist bias, an important selection effect popularized by the Swedish astronomer Gunnar Malmquist (1893–1982).[17] The effect must be understood and corrected for, which is complicated by the fact that there are also evolutionary effects that modify the underlying population. Distant objects are seen at a time the universe and these objects were younger. Furthermore, as objects evolve with time, their population representation changes over different periods of time.

For extended objects, i.e. those that are not as point-like as stars, the surface brightness can introduce a biasing trend. The night sky has an intrinsic brightness, partly due to the fluorescence of the Earth's atmosphere, but also as a result of the general stellar background. It is easier to detect objects of high rather than low surface brightness. If you examine Plate 6.3 closely, you will find dozens of faint galaxies, which are barely visible. This greatly affects the detection and measurement of extended faint objects, either regular nebulae in our own galaxy or galaxies of low surface brightness in the extragalactic world. The knowledge that these objects exist and the need to assess and count them galvanize efforts to optimize the transmission and control of scattered light in optical systems. Observations of the nearby universe and the modeling of galaxy evolution can correct for this. However, it remains extremely difficult to detect the large populations of faint galaxies.

The Drive to Classify

Synoptic representation triggers another epistemological stage: a classification process that tries to combine identical or similar features of several objects to create an encompassing class of objects or extract an archetypal object. In his 1811 paper, William Herschel presented two plates to show the main nebular shapes (Fig. 2.4). It was an undifferentiated mix of what we now know to be galactic nebulae and external galaxies. Herschel saw all of them as one class of objects and he tried to sequence them.[18] He suggested that the degree of condensation of "nebulae" was related to gravity, thus determining nebular shapes. Perhaps more importantly, in the light of later developments, Herschel inferred that the shapes of

---

[15] J. E. Keeler, Note on a Cause of Differences Between Drawings and Photographs of Nebulae, *Publications of the Astronomical Society of the Pacific*, 1895, Vol. 7, pp. 279–282.

[16] The impact of filters and wavebands with regard to galaxy morphology and classification is discussed by the authors in R. J. Buta, H. G. Corwin Jr. and S. C. Odewahn, *The de Vaucouleurs Atlas of Galaxies*, Cambridge: Cambridge University Press, 2007.

[17] G. Malmquist, On Some Relations in Stellar Statistics, *Arkiv för Mathematik, Astronomi och Fysik*, 1922, Vol. 16, pp. 1–52.

[18] M. J. Way, Dismantling Hubble's Legacy? in *Origins of the Expanding Universe: 1912–1932*, Astronomical Society of the Pacific Conference Series, 2013, Vol. 471, pp. 102–103.

"nebulae" were physically meaningful as criteria to be used by classifiers. Stephen Alexander also sought such an ordering (see Fig. 2.11 for a small collection of shapes assembled by Alexander).

Almost a century after Herschel, the German astronomer Max Wolf (1863–1932) of Heidelberg Observatory, also a pioneer of astrophotography and wide-field photography, introduced a classification scheme for "small nebulae" (*kleine Nebelflecken*) (Fig. 4.4). Having commented on the limited power of spectroscopy to sort nebulae – this had been attempted by American astronomer Edward Charles Pickering (1846–1919)[19] – Wolf used morphology as a criterion for classifying all "nebulae." Avoiding any inference about the intrinsic nature of the "nebulae," he wrote cautiously: "You will therefore need to be purely descriptive, without any consideration of a hypothesis about the origin of the mist, or the nature of the spectrum."[20] Wolf's drawings, derived from photographs, appeared to be of poor quality compared to those produced at Birr Castle 50 years before. While William Parsons' team depicted specific objects as accurately as possible, Wolf purposely mixed and averaged features. What mattered was not the accuracy of the sketch but the common structural information that the morphology conveyed.

In putting together his global scheme, Wolf extracted 23 "classes" that summarized the main shapes of the foggy nebular images. He rightly emphasized the dependence of the quality and shapes of the images delivered on the optical instruments used. Like Herschel's 1811 plates, Wolf's set is a mix of objects that included Milky Way nebulae and galaxies. Although spiral shapes and edge-on systems were clearly recognized, Wolf did not add much to the work of Parsons. Wolf's merit resided in his teleological purpose: to show that the natural sequence of shapes betrays deeper physical phenomena and processes. We will return to the topic of galaxy classification in Chapter 9.

High-Level Representations

As more quantitative physical measurements grew, the use of synoptic representations in nebular and galactic research increased. With the advent of photography, quantitative photometry became possible and the technique was quickly applied to images of galaxies; images or charts in different wavebands were calibrated and constructed.

Isophotal maps, connecting portions of equal brightness in the image of an object, turned out to be essential for comparing the same objects imaged at different wavelengths, especially for those obtained in the non-visible domains. It was particularly important to relate these findings with observations and images in the radio and the X-ray domains that astronomers had started working with during the second half of the twentieth century. Omar Nasim has given a fine description of the work of the young American astronomer Ebenezer Mason, who was introduced at the start of this chapter. Mason was probably the first to use

---

[19] Pickering led a large effort at Harvard College Observatory that resulted in an effective and successful spectral classification system of stars.
[20] M. Wolf, Die Klassifizierung der kleinen Nebelflecken, *Publikationen des Astrophysikalischen Instituts Königstuhl-Heidelberg*, 1908, Vol. 3, pp. 109–112 (translation by the author).

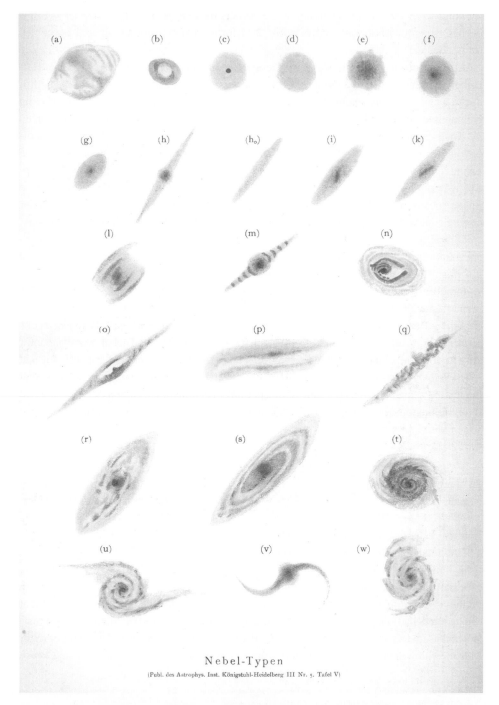

Fig. 4.4 Early classification of "nebulae" by Max Wolf. From Wolf (1908), *Publikationen des Astrophysikalischen Instituts Königstuhl-Heidelberg*. Courtesy of John G. Wolbach Library, Harvard College Observatory. 2016.

the method of lines of equal brightness in astronomical observation.[21] "It was first suggested by the method usually adopted for the representation of heights above the sea-level on geographic maps, by drawing curves which represent horizontal sections of a hill and valley at successive elevations."[22] In a parallel approach, John Herschel produced descriptive maps where a nebular drawing was started by making an accurate reference grid of stars, which served as a sort of triangulation basis; he then drew nebular intensity, carefully linking areas of equal brightness.

Isocontour maps of physical properties became the standard means of non-homomorphic representation. Nasim refers to the "isoline craze" that embraced the natural sciences during the nineteenth century.[23] Astronomy was one of these "infected" sciences. Isophotal maps as Mason invented them became extensively used in astronomy and continue to be a universal way to present imaging data (see Plate 4.1; Plate 7.2).

Digital imaging enables an additional step, that of creating intensity ratios of images of different wavelengths. Another dazzling example has been the use of isovelocity maps: the joining up of locations of equal radial velocity points across an object (Plate 4.2). Such maps have been used to establish the behavior of the orbital velocities of stars or gas as a function of the distance to a galaxy center. These rotational curves and maps have become essential tools for deriving the mass distribution within individual galaxies and their total mass. The use of "false colours" is most effective in highlighting the information to be conveyed.

Probably one of the most stunning non-homomorphic maps in modern astronomy is that of the cosmic microwave background (CMB), a relic of the distribution of matter in the universe as it was about 380,000 years after the Big Bang (Chapter 5). Since the serendipitous discovery by Arno Penzias and Robert Wilson in 1964, we know that the universe is filled with microwave radiation. The intensity distribution across wavelengths corresponds to a perfect black body at 2.725 K. The fossil glow peaks in the millimeter–centimeter wavelength range of the spectrum, the microwave region. The present-day CMB is the remnant light from the time when the mean temperature of the universe had cooled enough, due to expansion, for the free electrons to be captured by the protons. The universe became transparent and the photons were left to cool as the universe continued its expansion.

There were no stars or galaxies this far back in time. Intriguingly, the microwave background is not absolutely uniform. It is spotted with tiny fluctuations in temperature (Plate 4.3). These are of the order of only 1/100,000 K but have been precisely measured with very sensitive radio telescopes in space and on the ground. When mapped over the surface of the sky, the fluctuations in temperature show an orderly angular structure; the mottling has a pattern. The most straightforward interpretation of the pattern is that these structures correspond to regions of the young universe where matter had started to assemble under the

---

[21] J. Shears, et al., In the Footsteps of Ebenezer Porter Mason and His Nebulae, 2014arXiv:1401.7960. See also O. W. Nasim, *Observing by Hand: Sketching the Nebulae in the Nineteenth Century*, Chicago: University of Chicago Press, 2013, pp. 131–137.

[22] Cited in J. Shears et al., In the Footsteps of Ebenezer Porter Mason and His Nebulae, 2014arXiv:1401.7960, p. 3.

[23] O. W. Nasim, *Observing by Hand: Sketching the Nebulae in the Nineteenth Century*, Chicago: University of Chicago Press, 2013, p. 131.

action of gravity. As these enhanced structures grew, they led to large-scale structures and to the superclusters of galaxies we observe in the universe of today. The map of the spatial temperature fluctuations of the CMB is a powerful non-homomorphic representation of the distribution of matter in the young universe a few hundred thousand years after the Big Bang.

These few examples show that images as representations of nature do not need to be an exact replica of what an omnipotent human eye would see, also assisted by an instrument like a telescope or a microscope. Non-homomorphic images can convey information of a higher level.

This chapter concludes the overview of the different approaches and techniques that astronomers have developed over the centuries for viewing, sketching and photographing "nebulae," in brief, to make images of the "nebulae." In the next four chapters, the style changes. I present the world of galaxies using a more thematic approach, highlighting key individuals and techniques, as well as the specificities of imaging in each wavelength domain (radio, infrared, visible, ultraviolet, X-rays and gamma-rays) that have all helped to elucidate the surprising objects and phenomena of the extragalactic universe.

Part II

Images as Galaxy Discovery Engines

5

The One-Thousand-Year Journey

What are galaxies? No one knew before 1900. Very few people knew in 1920. All astronomers knew after 1924.

Allan Sandage[1]

The history of scientific discovery affords many instances where men with some strange gift of intuition have looked ahead from meager data, and have glimpsed or guessed truths which have been fully verified only after the lapse of decades or centuries.

Heber D. Curtis[2]

Few discoverers, if any, tried more carefully to avoid the most fundamental aspect of their findings than did Hubble.

Stanley Jaki[3]

Why Did It Take One Thousand Years to Figure Out What "Nebulae" Were?

"The best astronomer hardly known in those days," is how Allan Sandage described his fellow countryman, American astronomer Vesto Slipher (1875–1969).[4] Slipher was an outstanding spectroscopist. He measured the radial velocities of dozens of spiral "nebulae" and found that most were speeding away from us at speeds much higher than the escape velocity of the Milky Way. His finding was instrumental to the discovery of the expanding universe. Slipher had joined the Lowell Observatory in 1909, where the eccentric Percival Lowell, searching for life on Mars, had built a private observatory equipped with a 24-inch refractor (Fig. 5.1). Lowell, strong adherent of the Nebular Hypothesis, believed spiral "nebulae" were solar systems in the process of formation. He had hired Slipher to study planets. Starting around 1912, and adding to these planet studies, Slipher developed a

---

[1] A. R. Sandage, *The Hubble Atlas of Galaxies*, Washington: Carnegie Institution of Washington, 1961, p. 1.
[2] H. D. Curtis, Modern Theories of the Spiral Nebulae, *Journal of the Royal Astronomical Society of Canada*, 1920, Vol. XIV, pp. 317–327.
[3] S. L. Jaki, *The Milky Way, An Elusive Road for Science*, New York: Science History Publications, 1972, p. 302.
[4] A. R. Sandage, *Centennial History of the Carnegie Institution of Washington, Volume 1: The Mount Wilson Observatory*, Cambridge: Cambridge University Press, 2004, p. 450.

Fig. 5.1 Vesto Slipher in 1932. Credit: University of Chicago Photographic Archive, [apf6–04298], Special Collections Research Center, University of Chicago Library.

spectroscopic research program on "nebulae." Of independent mind, he slowly came to regard these objects as genuine galaxies.[5]

Slipher was totally oblivious to fame. As Sandage writes further, "Slipher's desire for astronomical recognition approached zero." Slipher later became mayor of the growing nearby city of Flagstaff, Arizona, and was in control of the main utility departments. He and his brother Earl C. Slipher were successful businessmen and rangers in the developing wild southwest. As the booming town sprawled and city lighting grew, Slipher was able to turn off the city lights when needed for the observation of faint objects, simply by activating a main switch he had installed in his observatory office.

The Gradual Dawning of Galaxies and the Law of Eponymy

The previous chapters have explored the various ways of making images of "nebulae." By examining the various means of "seeing things," e.g. observing with the unassisted eye,

---

[5] M. J. Way and D. Hunter (editors), *Origins of the Expanding Universe: 1912–1932*, Astronomical Society of the Pacific Conference Series, 2013, Vol. 471. A goal of the conference was to re-establish Slipher's recognition.

Fig. 5.2 Harlow Shapley on a radio broadcast at CBS in the 1940s. Photograph of Harlow Shapley, July 23, 1946. HUG 4773.80P (Box 2), Harvard University Archives.

hand sketching and photographing re-enacted the slow recognition that our gigantic stellar system is only one island in a colossal cosmic archipelago of billions of others. Establishing the Sun's position as a tiny member of the Milky Way itself has also followed a long and twisted path.

In contrast to some other scientific controversies, the realization that spirals are other stellar systems like our own Milky Way, located millions and even billions of light-years away, had limited public impact. There was little response from broad society, certainly nothing like the uproar that greeted Darwin's theory of evolution, Freudianism or the fictitious canals on Mars. There were reasons for the tame reaction. First, the "island-universe" view was not socially revolutionary. It did not disturb the order of things, even less the perceived course of events on Earth. Galaxies were not part of the scriptures or any other holy text. Secondly, the sequence of breakthroughs in physics, biology and psychology around the turn of the century may have acted as an intellectual desensitizer. Finally, compared with the practitioners of the nineteenth century, scientists of the twentieth century had become more specialized, more focused on one discipline; they often ignored or failed to participate in the major debates and controversies in fields other than their own. Naturalists or generalists of the eighteenth and nineteenth centuries who engaged in every intellectual debate were becoming an extinct group.

Even among astronomers, the subject appeared to raise more curiosity than deep interest. The so called "Great Debate" between Harlow Shapley (Fig. 5.2) and Heber Curtis at the meeting of the National Academy of Sciences in Washington, DC, on 26 April 1920 was a subdued affair, where a young, ambitious but insecure Shapley seemed to avoid any hard confrontation with the experienced Curtis. Reading the slides shown by Curtis, one realizes how structured and assertive Curtis was at conveying his perspective of spirals being external systems, hundreds of thousands of analogues of the Milky Way.

Shapley was skirting with the issue. Instead of tackling the nature and location of spirals, he tried to promote his arguments in favor of a "super Milky Way," 300,000 light-years in size, which embraced everything. Sarcastically, Shapley wrote privately: "The Andromeda Nebula, a giant among the external galaxies, is apparently not larger than the star cloud in Sagittarius and Scorpio which appears to form the nucleus of our Milky Way."[6] Referring to spirals, he preferred "to believe that they are not composed of stars at all, but are truly nebulous objects." Shapley was putting undue weight on nebular spectra that showed the emission lines typical of diffuse gaseous substances.

Curtis (see Fig. 3.4) took this spectroscopic argument and skillfully turned it on its head, stating that the spectra of spirals were "as would be expected from a vast congeries of stars." Five years before Hubble determined direct distances to spirals, Curtis surmised that the "great nebula" in Andromeda was at 500,000 light-years from us, and the more remote spirals at 10 million or more light-years. Michael Hoskin has provided a wonderful and detailed overview of the debate and related events, by drawing on archival material. He warned that some historians have erred in dramatizing a debate that really never took place, creating "an historical romance."[7]

Many excellent books have told the story of the discovery of the world of galaxies, emphasizing several fascinating aspects; the reader should refer to these works for a more complete analysis and other perspectives.[8] The same story seems to be retold each time with its variable and particular set of details. It is an intricate but wonderful story, and like kids at bedtime, we like to hear it over and over again. Here I will be brief. This chapter re-assembles in a more linear way some parts of previous chapters. There is a review of the battle to resolve "nebulae" into stars, a long and unsuccessful crusade, which at the same time helped in the mission to locate better telescopes at more favorable observing sites. The roles of key players in the discovery of galaxies are highlighted, and how this led to the dramatic follow-up finding, that the universe is expanding; and I consider how images drove the development of new ideas.

If a particular lesson is to be remembered, it is the intricacy of the scientific discovery process. Stigler's law of eponymy encapsulates this: no scientific discovery is named after its discoverer.[9] Stephen Stigler, a professor of statistics at the University of Chicago, demonstrated his paradoxical point in a straightforward manner: it was science historian

[6] Memo from H. Shapley to J. C. Duncan, 29 January, 1930, cited by O. Gingerich, Through Rugged Ways to the Galaxies, *Journal for the History of Astronomy*, 1990, Vol. 21, p. 79.

[7] M. Hoskin, The 'Great Debate': What Really Happened, *Journal for the History of Astronomy*, 1976, Vol. 7, pp. 169–182.

[8] Among several books, I note many excellent ones: S. L. Jaki, *The Milky Way – An Elusive Road for Science*, New York: Science History Publications, 1972; R. Berendzen, R. Hart and D. Seeley, *Man Discovers the Galaxies*, New York: Science History Publications, 1976; R. W. Smith, *The Expanding Universe – Astronomy's 'Great Debate' 1900–1931*, Cambridge: Cambridge University Press, 1982; H. S. Kragh, *Conceptions of Cosmos – From Myth to the Accelerating Universe, A History of Cosmology*, Oxford: Oxford University Press, 2007; M. Bartusiak, *The Day We Found the Universe*, New York: Vintage Books, 2010; A. Hirshfeld, *Starlight Detectives: How Astronomers, Inventors, and Eccentrics Discovered the Modern Universe*, New York: Bellevue Literary Press, 2014. For a concise overview, see J. D. Fernie, The Historical Quest for the Nature of the Spiral Nebulae, *Publications of the Astronomical Society of the Pacific*, 1970, Vol. 82, pp. 1189–1230.

[9] S. M. Stigler, Stigler's Law of Eponymy, *Transactions of the New York Academy of Sciences*, 1980, Vol. 38, pp. 147–158.

Robert K. Merton who had initially proposed the "law," not him. The law of eponymy is not a law in the scientific or logical sense, but a sociological statement about discovery.

Imagining "Nebulae"

As has been shown in the previous chapters, the stars of the celestial sphere have been observed, catalogued and mapped for several thousand years. Nebulous objects were found from time to time among the fixed stars as they moved rapidly across the sky; they were comets. However, a few diffuse objects appeared fixed against the sidereal vault; they became known as "nebulae." Chapter 1 gives an account of how the Persian scholar Abd al-Rahman ibn Umar al-Sufi first identified, in the year 964, what we now know to be a galaxy: the "little cloud" in the constellation of Andromeda. There then followed a silence of almost seven centuries. In 1614, Marius observed the "little cloud" with a telescope and gave a vivid description of the "nebula" in Andromeda as "the light of a candle, seen at some distance, shining through horn."[10]

We had to wait more than a century and a half for the exploration of the nebular world to take on a new life, with a more rigorous approach. The study of the nebular phenomenon followed two parallel tracks: first, an empirical one, solidly observational, mainly by the German-born musician and dilettante astronomer William Herschel and his astute sister Caroline; and secondly, a highly speculative avenue, roamed by three theological and philosophical writers, Thomas Wright, Johann Heinrich Lambert and Immanuel Kant.

Although being of cautious mind, William Herschel tried to make sense of the numerous objects scattered across most of the sky. Like others, he battled with the challenge of some "nebulae" being, or appearing to be, resolved into stars and others not, caught in the midst of many contradicting reports. Herschel made recurrent hypotheses and went through paradigm shifts: as we saw, he first surmised that "nebulae" could simply be clusters of unresolved stars, due to their large distances, other "milky ways"; later, he appeared to favor nebular matter of an unknown state, some shining fluid that existed between the stars; later still, he saw "nebulae" as solar systems in formation. Although he changed his mind regarding their nature, "William did more than any other astronomer to trigger the transformation from the eternal, clockwork universe of Isaac Newton and Gottfried Leibniz to that of modern astronomy in which everything from individual stars to the cosmos itself changes over time."[11]

Herschel had been neither alone nor the first on the speculative path to interpret "nebulae." Their mystery had already captured the attention of eighteenth-century philosophers. With an increased vigor, the most eminent thinkers, Wright, Lambert and Kant, undertook to describe the large-scale structure of the heavens. They employed a mix of theosophical and physical considerations.

---

[10] Cited in G. P. Bond, An Account of the Nebula in Andromeda, *Memoirs of the American Academy of Arts & Sciences*, 1848, Vol. 3, p. 76.

[11] M. Hoskin, Caroline Herschel as Observer, *Journal for the History of Astronomy*, 2005, Vol. XXXVI, p. 396.

The first, Thomas Wright of Durham (1711–1786), remains known for his highly speculative and moral work, *A Theory of the Universe*, published in 1734. He dealt with the various astronomical phenomena and sought a unified explanation. He tried to explain novae, variable stars, nebulae, comets and the Milky Way in terms of volcanoes in the sky.[12] Wright had little impact on the thinking of contemporary astronomers. William Herschel had a copy of Wright's book and annotated it. As commented by Michael Hoskin, Herschel never expected to learn anything from the work.

Johann Heinrich Lambert (1728–1777) was the second moral "cosmologist" to be captivated by "nebulae." He elaborated on a hierarchical system of clusters of stars spreading across millions of light-years of sidereal space. William Herschel had no sympathy either for the "fantastic imaginations" of Lambert. It was an issue of character: an opposition between the hard, reluctant empiricist that Herschel was and dreamers of subjective moral order.

Wright's ideas on trying to integrate the moral and scientific picture of the universe did influence fertile minds, including that of Immanuel Kant; Kant was less restrained by theological references and, fortunately, he wrote more clearly than Wright. Using a consistent argument and borrowing on Newton's theory of gravitation, Kant's speculative approach in explaining cosmic shapes showed good insight. He refuted Pierre Louis Maupertuis' idea that nebulae were huge flattened stars induced by their rapid rotation. Instead, Kant proposed a drastically different concept, "that there are no such individual huge stars but systems of many stars, whose distance makes them appear in such a narrow space, that the light, which cannot be seen for each individual star, because of the countless crowd of them, comes out in a uniform pale glow."[13]

It is remarkable that shape or morphology played a significant role in Kant's interpretation and his model for "nebulae." Before it became a contentious debate among astronomers, Kant became known as the champion of the "island-universe" concept: most "nebulae" are distant Milky Way-like systems. However, he could neither prove this, nor was he equipped to do it.

The Nebular Hypothesis

In proposing, in the same little book, a precursor view of the nebular hypothesis, Kant threw a snag onto the elusive path of unraveling the nature of "nebulae." It was a concept that captivated, often distracted, many astronomers of later generations and acted as a double-edged sword. Why? Because the nebular shapes that were revealed by telescopic drawings and early photographs showed patterns consistent with the mechanism Kant had proposed to describe the formation of the Sun, its planets and their satellites. Much of the

---

[12] M. Hoskin, The Cosmology of Thomas Wright of Durham, in *Stellar Astronomy: Historical Studies*, Chalfont St. Giles: Science History Publications, 1982, pp. 113–114.

[13] I. Kant, *Universal Natural History and Theory of the Heavens* (translated by Ian Johnston), Arlington: Richer Resources Publications, 2008, p. 37.

controversy revolved around images of "nebulae" and their physical meaning.[14] Many images of "nebulae" were loosely viewed as flattened whirlpools of condensing cosmic matter. Furthermore, if they were proto-planetary systems, "nebulae" could only be relatively nearby, thus part of the Milky Way, and not the island-universes Kant hypothesized about.

Here is how Kant presented his hypothesis of an interstellar cloud of matter collapsing under the action of gravity.

In the greatly expanded space in which the spread out elementary basic material prepares developments and systematic movements, the planets and comets are built up only out of those parts of the elementary basic matter moving downward towards the central point of the force of attraction, which, through their fall and the reciprocal interaction of the particles collectively, were precisely adjusted for the velocity and direction required for orbital motion.... Now, because these lighter and volatile parts are also the most effective at maintaining a fire, we see that, with their addition the body at the central point of the system has the distinction of becoming a flaming sphere, in a word, a sun. By contrast, the heavier and inert materials and those particles which are poor fuel for a fire will make planets which are robbed of these properties merely cold and dead clusters.[15]

I quote the extensive passage because many researchers and the "grand amateurs" of the nineteenth and early twentieth centuries were bewitched by this impressive image of the mind and imposed it on the "nebulae" they observed, sketched and photographed. The Nebular Hypothesis gained strength and popularity when, in 1796, Pierre Simon Laplace presented a rigorous mathematical concept in *Exposition du système du monde*.[16]

On the Road of Confusion

The observations conducted at Birr Castle constituted an observing program in the modern sense: first, a systematic search and examination of the objects found by Herschel and catalogued in the General Catalogue of nebulae and clusters. Because of the controversy surrounding the reality of several objects, the first step of the work was to confirm the existence of these objects; a few were found not to exist, to have been incorrectly located or to be duplicates. Beyond the verification process, the Birr Castle observers paid meticulous attention to the description of the objects and to drawing them. Importantly, William Parsons' initial motivation was to resolve the "nebulae" into stars with the Leviathan; the plan was soon derailed with the discovery of spirality in dozens of them.

In his long summary article of 1878, Lawrence, the Fourth Earl of Rosse, Parsons' son, never even hinted at the nature of these "nebulae." The reporting was strictly factual. One

---

[14] S. Schaffer, On Astronomical Drawing, in *Picturing Science, Producing Art*, C. A. Jones and P. Galison (editors), New York: Routledge, 1998, pp. 441–474.

[15] I. Kant, *Universal Natural History and Theory of the Heavens* (translated by Ian Johnston), Arlington: Richer Resources Publications, 2008, pp. 105–106.

[16] P. S. Laplace, *Exposition du système du monde*, Paris: Imprimerie du Cercle-Social, 1796.

can, however, imagine the hours of discussion and speculation among the members of the close-knit team. Outside, the "nebular" controversy and the confusion about what they were, unresolved groupings of stars, some sort of shining fluid or such fluid mixed with stars, continued to rage. Paradoxically, as imaging techniques became more sensitive, with larger and better telescopes and the increased use of the photographic plates, the confusion seemed to be amplified: some "nebulae" were shown or claimed to be resolved into stars, while most remain "cloudy." Proponents of the "local" hypothesis argued that if "nebulae" were resolved into stars, they had to be relatively nearby, that is part of the Milky Way system. The confusion of the end of the nineteenth century spilled over into the first decades of the twentieth century, as swings of opinion favored first the local hypothesis, then the universe-islands view, back and forth.

Resolving "nebulae" into stars became a battleground. The "resolution" craze was stimulated by claims by the Fourth Earl of Rosse and George Bond that parts of the Orion Nebulae split into numerous stars. "Resolving stars" reinforced the century-old argument put forward first by Galileo: with sufficient optical power, all nebular forms can be resolved into individual stars. Robert Smith quotes a triumphant passage from George Bond's September 1847 observing notes on the Orion Nebula. An excited Bond, happy to show the power of the new Harvard College 15-inch refractor, exclaimed: "Resolved. Mottled. Abundance of Stars."[17]

Claims of resolution of other "nebulae" into individual stars were heard repeatedly. Both proponents and opponents of the extragalactic hypothesis used the partly fallacious argument of resolution to support either the local or external version of the nature of "nebulae." Current insights on how so many claimed to have resolved individual stars suggest that the optics of the telescope used, especially metallic mirrors, combined with variable "seeing," could have generated an apparent granularity to the image as viewed in the eyepiece of these instruments. Stated rather bluntly by the Pulkovo Observatory director Otto Wilhelm Struve, "the alleged miracles of resolution are nothing but illusions." He was right. We have here a case of images turning out to be unreliable conveyors of reality.

At the top of the pile of nebular puzzles, observers claimed to have detected a variability of brightness in a few nebulae, in some instances fading over time. Although there were a few cases of true nebular variability, due to erupting dust-embedded protostars, most of these were bogus. In particular, the several claims of the variability of the Orion Nebula were all erroneous. Nevertheless, changes in apparent brightness were used as a strong argument in favor of the local hypothesis. The argument, and it was logical, was that no giant systems of stars could cohesively vary in unison. Simon Schaffer has recounted this fascinating story, highlighting the debate among astronomers surrounding the elusive perception of changes.[18]

---

[17] R. W. Smith, *The Expanding Universe, Astronomy's Great Debate 1900–1931*, Cambridge: Cambridge University Press, 1982, p. 45

[18] S. Schaffer, On Astronomical Drawing, in *Picturing Science, Producing Art*, editors C. A. Jones and P. Galison, New York: Routledge, 1998, pp. 441–474.

The Arrival of Spectroscopy

A transformational technique was introduced during the first decades of the nineteenth century: spectroscopy was employed to spread the light of stars and "nebulae" as a function of wavelength and to analyze it.[19] German physicist and optician Joseph von Fraunhofer (1787–1826) had invented the spectroscope in 1814. He explored the light from the flames of several substances and analyzed sunlight, discovering 574 "dark lines" in its spectrum. Another "grand amateur," British astronomer William Huggins (1824–1910) and his wife Margaret Lindsay Huggins (1848–1915), pioneered stellar spectroscopy. They found that the spectra of celestial objects, e.g. positions or wavelengths of the various "dark lines," resembled those of terrestrial substances; they concluded that stars contain some of the same elements found on Earth.

Applying spectroscopy to the "nebulae" was difficult as the objects were extended and very faint. Better optical designs finally made the avenue fruitful, and more telescopes were equipped with spectrographs, and "nebulae" became prime targets for spectroscopists. What the nebular spectra initially showed did not rally viewpoints but strengthened divergence. In 1864, William Huggins obtained a spectrum of a planetary nebula in the constellation of Draco. It was totally different from the spectra of the stars (Plate 5.1). A single strong emission line dominated the spectrum, a clear indication that some sort of hot tenuous gas was the emitting source. Also differing from the Sun and stars that produced a continuous range of colours broken by "dark lines," all spectacular nebulae, like the Orion Nebula, showed monochromatic emission: the light, dispersed by the spectroscope, was concentrated over a very narrow range of wavelengths – called "emission lines," as they were bright, colourful features, the reverse of the dark lines seen in the solar or stellar spectra. Observing other "nebulae," it became obvious that many of them appeared to be made of a gaseous and fluorescent substance. Huggins wrote with assurance "The riddle of the nebulae was solved. . . . the light of this nebula had clearly been emitted by a luminous gas."[20] No one knew then that stars were also gaseous but of much higher density than nebulae.

The first spectroscopy results, which showed "nebulae" to be some sort of shining fluid, appeared to support the local hypothesis. However, Huggins and Margaret Lindsay soon found that "nebulae" displayed a range of spectral properties, with many others displaying instead continuous spectra, similar to those of the Sun and of other stars. Hence, in the end, they remained cautious about the nature of "nebulae." It was wise. Too many were committing again and again the sin of putting everything in the same bag.

How the Milky Way Became the Universe – For a While

William Herschel had already made the observation that the band of the Milky Way was relatively devoid of "nebulae," while large portions of the sky above and below the plane

[19] J. B. Hearnshaw, *The Analysis of Starlight: One Hundred and Fifty Years of Astronomical Spectroscopy*, Cambridge: Cambridge University Press, 1986.
[20] W. Huggins and Lady M. Huggins, *The Scientific Papers of Sir William Huggins*, London: Wesley & Son, 1909, p. 106.

of the Milky Way were sprinkled with them. In the stubborn tradition of seeing humankind at the center of all things, the apparent symmetry in the distribution of "nebulae" about the Milky Way plane was hailed as an irrefutable argument for the local hypothesis. Nebulae had to belong to the Milky Way in order to display such a symmetrical distribution. Favoring this argument, English writer Richard Proctor (1837–1888) posited that this "zone of avoidance" was an indication that spirals were part of the Milky Way: How could the "nebulae" assemble this way around if they were totally independent of it? How would the outer worlds know we were here if they were scattered across huge distances?

This galactocentric view gained support in the 1910s, when Shapley noted that globular clusters of stars also obeyed the exclusion from the avoidance zone (Chapter 4). Globular clusters are giant, tight assemblies of hundreds of thousands, even millions, of stars in systems of spherical shape. Shapley's argument was that the zone of avoidance corresponded to the high density of stars. Consequently, globular clusters went missing because they were torn apart by tidal forces when they entered the region and were destroyed. A convinced Milky Way centric, Shapley extended the same argument of destruction to spirals, reinforcing his opinion that spirals were members of our galaxy. His inference for a super Milky Way was based on the assumption that sidereal space is completely transparent: there is nothing to block or absorb starlight as it travels thousands of light-years of interstellar space. Shapley's explanation for the missing spirals in the band of the Milky Way was completely false. At that time there was little recognition that interstellar dust was severely blocking our view in several directions of the sky.

As the nineteenth century came to a close, the resolution of "nebulae" into stars, the spectroscopic evidence of a gaseous composition and the discovery of variations in nebular brightness appeared to carry the day for spirals, and hence all "nebulae," to be local. The enthusiasm for the local view grew in strength and boldness. Our old friend Isaac Roberts encouraged the supporters of the Nebular Hypothesis to study his unique photographs of the Andromeda Nebula "to see a new solar system in process of condensation from the nebula." No less enthusiastically, he envisioned the two small galaxy companions of M31 as unique laboratories for cosmogony, as "the two [smaller] nebulae seem as though they were already undergoing their transformation into planets."[21] Fanciful speculation was again getting ahead of hard facts.

An unexpected phenomenon seemed to lend weight to "nebulae" being part of the Milky Way. The "great nova" of 1885 in Andromeda became as bright as a tenth of the whole Andromeda Nebula; most astronomers thought it inconceivable that a single star could have outshone millions of stars, as would be the case for a distant object that constituted a multitude of stars. This puzzling new star just added to the quandary, and suggested that the Andromeda Nebula had to be small and nearby, as argued by Shapley during the "Great Debate." We know today that S Andromedae or SN 1885 was not a nova but a supernova, the first ever noted outside the Milky Way. As seen in Chapter 3,

---

[21] I. Roberts, Photographs of the Nebulae in M32, h44, and h51 Andromedae, and M27 Vulpeculae, *Monthly Notices of the Royal Astronomical Society*, Vol. 49, p. 65.

novae were well known because of their high frequency. Supernovae are much less frequent events as they involve massive stars, which are sparser in population. When a massive star explodes, it releases, over a few weeks, as much light and mechanical energy as the star ever did over its lifetime of millions of years. The luminosity of a supernova can thus outshine an entire galaxy. This difference between novae and supernovae came to be understood a few decades later, following the work of Walter Baade and Fritz Zwicky in 1934.[22]

Robert Smith has nicely summarized the conflicting views and the triumphal mood of the tenants of the local hypothesis at the turn of the century. "By the late 1880s, then, an astronomer could exploit the bright line spectrum of some nebulae, the peculiar distribution of nebulae, the photographs of the Andromeda Nebula and the 1885 'nova' to form the seemingly overwhelming case against the island universe."[23]

But things were about to change swiftly and dramatically in favor of the island-universe concept. The early decades of the twentieth century saw the spectacular rebirth of the externality of "nebulae" and the establishment of a world of galaxies. Oddly, the triumphal change came first from spectroscopy, then later from imaging techniques that finally allowed the measurement of accurate distances to the "nebulae."

Moving "Nebulae" Out of the Milky Way

In 1899, the German astronomer Julius Scheiner (1858–1913) of the Potsdam Observatory presented his spectroscopic results on the Andromeda Nebula in a succinct two-page paper. "No traces of nebular lines are present, so that the interstellar space in the Andromeda nebula, just as in our stellar system, is not appreciably occupied by gaseous matter."[24] Comparing features of Andromeda with our own Milky Way, Scheiner suggested that the irregularities and "streams" of the Milky Way could be quite well accounted for if they were regarded as a curl of "spirals." He stated with certainty that spiral nebulae were giant star clusters; the continuous spectrum of the decomposed nebular light was the synthesis of the light of a multitude of unresolved stars. The island-universe proponents finally had spectroscopy on their side. More spectroscopic evidence for the externality of most "nebulae" was to come a decade later. Scheiner was not impressed with Roberts' view that the Andromeda Nebula was a forming protoplanetary system.

Nevertheless, to have irrefutable proof of the externality of most "nebulae" it was necessary to overcome the barriers of determining accurate distances to them. Only this could settle the enduring debate. A crucial observational wave developed to study and understand variable stars. Until the twentieth century, variable stars were simply thought to be pairs

---

[22] W. Baade and F. Zwicky, On Super-Novae, *Proceedings of the National Academy of Sciences of the United States*, 1934, Vol. 20, pp. 254–259.

[23] R. W. Smith, *The Expanding Universe, Astronomy's Great Debate 1900–1931*, Cambridge: Cambridge University Press, 1982, p. 5.

[24] J. Scheiner, On the Spectrum of the Great Nebula in Andromeda, *The Astrophysical Journal*, 1899, Vol. 9, pp. 149–150. The spectrum of Messier 31 had required a 7.5-hour exposure.

of stars, i.e. double stars, eclipsing each other, with one star passing in front of the other and dimming its light. However, it was possible that the variability in brightness could have other causes. It took time for astronomers to realize and demonstrate that many stars underwent radial pulsations: that is periodic expansion and contraction of the whole star. A particularly important type of variable pointed towards the light at the end of the tunnel.

Exploding stars such as novae and supernovae are variable, but they are not pulsating. Even as irregular variables, novae can be used as "standard candles." These objects belong to the same class; they have a known brightness and variable behavior, which are a function of time (light curves), as based on direct measurements; for example, a nearby object of the same family whose distance has been established by the parallax method. In particular, they appeared to attain roughly the same maximum absolute magnitude. By comparing this known luminosity to a distant object's observed brightness, the distance of the latter can be derived using the inverse square law. Heber Curtis, George Ritchey and Knut Lundmark, a young Swedish astronomer who was then working in the United States, studied novae in "nebulae," as detailed in Chapter 3. The three astronomers published independently on novae they found in "nebulae." Novae were recognized in Messier 31 as clones of galactic ones. By 1917, Ritchey, Curtis and Lundmark had estimated the distances of those faint novae to be about 500,000 to 650,000 light-years, well outside the limit of even the super Milky Way of Shapley. These conclusions were based on the superb images provided by the refurbished 36-inch Crossley at the Lick Observatory and the new 60-inch on Mount Wilson.

Speeding "Nebulae"

With studies on novae paving the way, two other independent directions of study led to the determinations of distance that put most "nebulae" well outside the boundaries of the Milky Way. In a splendid paper published in 1917, Vesto Slipher reported on the large radial velocities he had derived from the spectroscopy of several "nebulae." Slipher's spectroscopic observations were the results of a painstaking and relentless program using the 24-inch refractor at the Lowell Observatory, Arizona. Slipher was as patient as he was quiet, consistently compiling multiple nights of exposure to get 20 to 40 hours of photons trickling onto each photographic plate. To obtain the velocities of the "nebulae," he used the Doppler effect, which shifts the position of spectral lines in proportion to their relative radial velocities, along the line of sight. He derived the movements of the "nebulae" with respect to the Milky Way. What Slipher found was phenomenal.

Although he measured a few objects like the Andromeda Nebula to be moving towards us, the majority (17 out of 21) were receding at high velocities, in some cases reaching 1,100 km/s: Slipher surmised most correctly that at such speeds, these objects could not be retained by the gravity of the Milky Way. At the end of his short note, he expressed his opinion: "It has for a long time been suggested that the spiral nebulae are stellar systems seen at great distances. This is the so-called 'island-universe' theory, which regards our stellar system and the Milky Way as a great spiral nebula that we see from within. This theory, it seems to me, gains favor in the present observations."[25]

Also leading the charge in distance determination, the brilliant Estonian astronomer Ernst Öpik (1893–1985) undertook a different and very original approach, the results of which he published in 1922 (Fig. 5.3). He was one of the most visionary astrophysicists of the first half of the twentieth century. Öpik analyzed the internal rotational motion of Messier 31, the "great nebula" in Andromeda, using spectroscopic data. He inferred its luminous mass from its surface brightness, assuming that the energy radiated by stars had the same behavior as in the Milky Way and that the velocities of stars in the system obeyed Newtonian dynamics. Employing simple and clever assumptions, he derived the distance to the Andromeda Nebula to be about 1.5 million light-years, his value being the closest of its day to the modern one of 2.6 million light-years. Öpik's method, known as the virial theorem (Chapter 8), is still in use today by researchers studying distant galaxies and galaxy clusters. Slipher and Öpik had confirmed the extragalactic scenario based on solid physical arguments. Now, let us return to variable stars.

How Henrietta Leavitt Gave the Key to Edwin Hubble

By the late eighteenth century, hundreds of regular variable stars had been identified and monitored. Harvard College Observatory, under its energetic directors Edward Charles Pickering (1848–1919) and his successor, Shapley, led the effort. Some variables were double stars, true eclipsing binary systems due to the orientation of their orbital plane with respect to us, but most were of a different nature. In the 1920s, as stellar interiors and atmospheres became better understood, regular variables were explained as being caused by radial pulsations: the star's volume and temperature change in a steady cycle of ups and downs, causing variable luminosity. This turned out to be especially true for a type of variable star called a Cepheid, the name coming from the first known representative in the constellation of Cepheus. The British amateurs Edward Pigott (1753–1825) and John Goodricke (1764–1786) discovered this special class of variables in 1784; that same year, Goodricke measured that δ Cephei varied by one magnitude over a period of 5.37 days. Sadly, the young Goodricke died from pneumonia, which he contracted during the numerous nights of observing variable stars in cold England; he was only 22.

Cepheid variable stars had thus been known for more than a century, but their observational properties were not fully understood. A detailed study of Cepheid variables by Harvard College Observatory astronomer Henrietta Swan Leavitt (1868–1921) changed everything regarding the astrophysical importance and use of these stars (Fig. 5.4).[26] Henrietta Leavitt worked as a "computer" under Edward Pickering at $10.50 per week. She was assigned to count images on photographic plates, and to make photometric measurements of variable stars.

---

[25] V. M. Slipher, Nebulae, *Proceedings of the American Philosophical Society*, 1917, Vol. 56, pp. 403–409.
[26] G. Johnson, *Miss Leavitt's Stars: The Untold Story of the Woman Who Discovered How to Measure the Universe*, New York: W. W. Norton & Company, 2005.

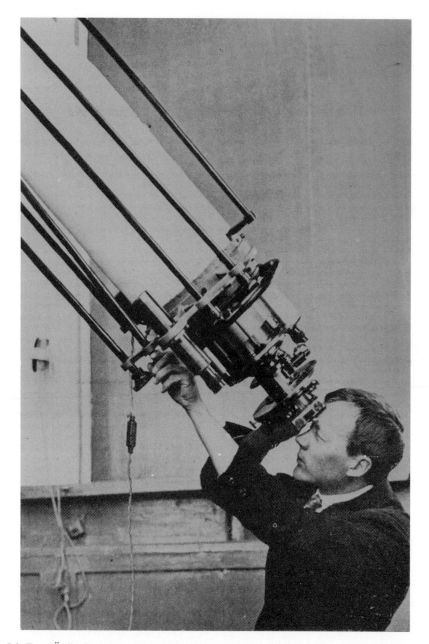

Fig. 5.3 Ernst Öpik observing with the 8-inch Zeiss refractor of Tartu Observatory, Estonia. Credit: Tartu Observatory (University of Tartu Museum Collections, UAMF 211_15) and Armagh Observatory and Planetarium.[27]

---

[27] This image is very similar to a photograph published in Tartu Tahetomi Kalendar XXI, 1944.

Fig. 5.4 Henrietta Leavitt. Courtesy of the American Association of Variable Star Observers (AAVSO).[28]

In 1908, Leavitt published her findings of the analysis of many photographic plates of the stars in the Magellanic Clouds, which had been obtained with the Bruce 24-inch telescope at Arequipa, Peru. Among 1,777 variables, she identified a class of variable stars that obeyed a surprisingly close relation between the luminosity and the period; while apparently fainter, they had similar features to the well-known galactic Cepheids.[29] The most luminous variable stars had the longest periods of variability, and the fainter one, the shortest. As these stars are on average very luminous, they could be seen to great distances and used as reliable standard candles. By simply measuring the period of a cycle of pulsation, the luminosity of the star could be derived in a straightforward manner (Fig. 5.5). It was magnificent work co-authored by Pickering.[30] Leavitt had provided the magic key to unlock the safe for measuring large cosmic distances. Unfortunately, Leavitt received little recognition in her time.

---

[28] Photograph of Henrietta Leavitt as it appears in *Popular Astronomy*, 1922, Vol. 30, p. 197.

[29] H. S. Leavitt, 1777 Variables in the Magellanic Clouds, *Annals of the Harvard College Observatory*, 1908, Vol. 60, pp. 87–108.

[30] H. S. Leavitt and E. C. Pickering, Periods of 25 Variable Stars in the Small Magellanic Cloud, *Harvard College Observatory Circular*, 1912, Vol. 173, pp. 1–3.

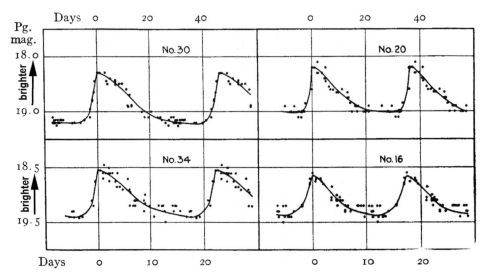

Fig. 5.5 Light curves of four Cepheid variables in the nearby galaxy Messier 33. From Hubble (1926), *Contributions from the Mount Wilson Observatory*.

Leavitt's work fundamentally changed the paradigm of distance determination by providing a very reliable ruler with which to measure the universe. Her work made huge impact. The game of the day became measuring distances using Cepheids as standard candles.[31] The triumphal act was for someone other than Leavitt. The American astronomer Edwin Powell Hubble (1889–1953) turned the Cepheid "key" in a majestic way and unlocked the distances to the galaxies. These were indeed staggering.

Hubble Cuts the Gordian Knot

Appointed at Mount Wilson Observatory in 1919, the young Hubble rapidly gained access to the newest and most powerful telescopes, the 60-inch and 100-inch of Mount Wilson Observatory (Chapter 3) (Fig. 5.6). The powerful telescopes were located on the dry mountains of the San Gabriel range just northeast of the city of Los Angeles. The night sky was then extremely dark and the stable atmosphere provided telescopic images of great definition. Hubble achieved his epoch-marking determination of extragalactic distances by using the newly calibrated variable stars, pulsating Cepheids. As stated by authors Harry Nussbaumer and Lydia Bieri, Hubble "cut the Gordian knot."[32]

Hubble's Ph.D. thesis topic had been the photographic investigations of faint nebulae based on work with the Yerkes Observatory 24-inch reflector, designed and built by Ritchey

[31] H. S. Leavitt, 1777 Variables in the Magellanic Clouds, *Annals of the Harvard College Observatory*, 1908, Vol. 60, pp. 87–108.
[32] H. Nussbaumer and L. Bieri, *Discovering the Expanding Universe*, Cambridge: Cambridge University Press, 2009.

Fig. 5.6 Edwin Powell Hubble scanning a photographic plate. Credit: Image courtesy of the Observatories of the Carnegie Institution for Science Collection at the Huntington Library, San Marino, California.

(Chapter 3). Returning from World War I work, Hubble started using the recently commissioned 100-inch Hooker reflector, initiating his formative program to observe "nebulae" in a systematic way. Following the works of Ritchey, Lundmark and Curtis, Hubble was expecting to find lots of new novae in the "nebulae" he was photographing. He obtained several photographic plates and compared them with archival ones, especially those acquired with the 60-inch, as this telescope had been in operation since December 1908.

Novae he found, but one peculiar object in Messier 31 caught his attention, as the same object was visible on plates taken a few years earlier (e.g. 1913). It could not be a nova since novae are only very rarely recurrent, and when they are, it is after periods of many years of dormancy. Initially, he erroneously marked the object as a nova with "N" for nova; he subsequently crossed it out with an X and correctly identified the star as a Cepheid variable with a period of 31.4 days (Fig. 5.7). Soon Hubble found several other Cepheids. Using Leavitt's period–luminosity relation, he derived a distance of 680,000 light-years to the Andromeda Nebula. He prepared his work for publication; it was near the end of 1924.

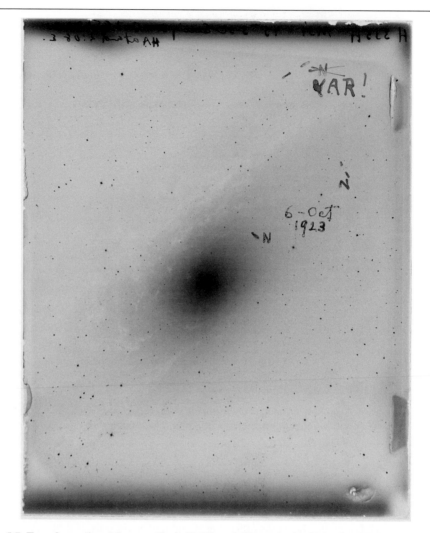

Fig. 5.7 **Transformational Image: Cepheid V1 and Novae in Andromeda (Messier 31).** Plate H335H of Messier 31 obtained by Edwin Hubble on the night of October 5–6, 1923. The letters N indicated novae. The top right N was crossed out and marked instead "VAR!". Hubble originally thought the star was a nova, but eventually discovered that it varied in brightness like a Cepheid. This is one of the most memorable images in the whole history of astronomy. Credit: Image courtesy of the Carnegie Observatories.

By then, Hubble had identified 36 variables in Messier 31; 12 of them were Cepheids (see Plate 5.2).

There are two interesting side anecdotes to the detection of Cepheids in "nebulae." Allan Sandage has recounted how a young observatory assistant, Milton Lassell Humason

Fig. 5.8 Milton Humason. Credit: Image courtesy of the Observatories of the Carnegie Institution for Science Collection at the Huntington Library, San Marino, California.

(1891–1972), who had started as a mule driver at Mount Wilson, became involved in the observatory scientific work (Fig. 5.8). Humason was doing photographic plate measurements for Harlow Shapley, who was then with the Mount Wilson Observatory. In the late 1910s, the curious Humason approached Shapley asking him if some of the variable stars he had noticed in "nebulae" could be Cepheids. An incredulous Shapley exclaimed: "Impossible, the nebulae are part of the big Milky Way."[33] For Shapley, nothing existed outside his Milky Way. Later, in 1924, in a letter to Hubble, who had informed him about the Cepheids he had found in Messier 31, the skeptical and almost satirical Shapley wrote back: "Your letter telling of the ... variable stars in the direction of the Andromeda nebula is the most entertaining piece of literature I have seen for a long time." As Owen Gingerich pointed out, Shapley did not say "in" the Andromeda nebula, but "in the direction of ..."[34] These stories, as several others, show how Shapley had a "talent" for missing the train, and as Walter

---

[33] Story told in A. R. Sandage, *Centennial History of the Carnegie Institution, Volume 1: The Mount Wilson Observatory*, Cambridge: Cambridge University Press, 2004, pp. 495–598.
[34] O. Gingerich, Through Rugged Ways to the Galaxies, *Journal for the History of Astronomy*, 1990, Vol. XXI, p. 78.

Baade harshly commented, to produce works that "never went beyond trivialities." Nevertheless, Shapley did produce some solid pieces of astronomical work. For many years, he remained a staunch supporter of the local hypothesis, seeing everything as part of a super Milky Way; this was in part due to the work of his friend at Mount Wilson, Adriaan van Maanen (1884–1946), a colleague of Hubble, who claimed – wrongly – to have measured the rotational motion of "nebulae" over short periods of time. If real, such motions would have implied velocities close to that of light. Van Maanen's measurements were erroneous and his interpretation fallacious.

For puzzling reasons, Hubble did not attend the 1924 December meeting of the American Astronomical Society to present his groundbreaking results. Instead, it was Henry Norris Russell who read a communication of Hubble's finding of "Cepheids in spiral nebulae" at the session of 1 January 1925. Historian Marcia Bartusiak described that moment emphatically as "the day we discovered the universe."[35] As has been described and emphasized throughout this book, the "day" had been a very long day.

Continuing his work with the 100-inch of Mount Wilson Observatory, Hubble identified and measured several more Cepheids in other "nebulae." From observations of Messier 33 and NGC 6822, he quickly derived accurate distances for these systems and published his results. "The publications of 1925 definitely settled and closed the debate on island-universes."[36] And Hubble started employing the descriptor "extragalactic nebulae." It was never clear if by "extragalactic" Hubble meant outside the Milky Way or simply above or below the plane of the Milky Way. It was only after Hubble's death in 1953 that the word "galaxies" began to be used broadly.

Having opened wide a giant field of research, Hubble also provided a lasting legacy by developing a classification scheme for galaxies based on their shapes. Combining earlier schemes, he proposed his own: " . . . nebulae are closely related members of a single family. They are constructed on a fundamental pattern that varies systematically through a limited range. The nebulae fall naturally into an ordered sequence of structural forms. . . . "[37] I come back to this topic in Chapters 9 and 10.

Galaxies and the Expanding Universe

Once distances to galaxies could be reliably established, several staggering discoveries quickly unfolded. Indeed, these measurements opened up a box of treasures. First, the universe was found to be expanding; this was an awesome aftermath, one of the greatest findings by the human mind in its quest to determine the nature of the universe.

Unknowingly and unwillingly at first, Albert Einstein (1879–1955) was the scout on the path to that stupendous discovery. In 1915 and 1916, Einstein had proposed and developed a new theory of gravitation that became his most famous legacy.[38] The theory of

[35] M. Bartusiak, *The Day We Found the Universe*, New York: Vintage Books, 2010.
[36] H. Nussbaumer and L. Bieri, *Discovering the Expanding Universe*, Cambridge: Cambridge University Press, 2009, p. 62.
[37] E. P. Hubble, *The Realm of the Nebulae*, New Haven: Yale University Press, 1936, p. 56.
[38] A. Einstein, Die Grundlage der allgemeinen Relativitätstheorie, *Annalen der Physik*, 1916, Vol. 49, pp. 769–822.

general relativity brought about a scientific revolution and a total change of paradigm with respect to Newtonian physics. Postulating the invariance of physical laws, Einstein surmised that space and time were not absolute. Daringly, he unified space, time and gravitation to describe a four-dimensional warping world, one with dips and ripples caused by concentrations of mass: stars, stellar clusters, galaxies and galaxy clusters. The "squashing" of time depended on the force of acceleration, and the curving of space hinged on the amount and concentration of mass.

John Archibald Wheeler (1911–2008) summarized the new Einsteinian physics most eloquently: "Spacetime tells matter how to move; matter tells spacetime how to curve."[39] Particles without mass, like photons, and particles of matter, either ordinary or dark, follow curved space. Everything in the universe had to obey a new geometry of spacetime. But there was a hidden potency in Einsteinian spacetime: it is dynamic, an astonishing property that Einstein did not want to see at first.

To Einstein's surprise and dismay, the equations describing the new universe also indicated that the new spacetime he had invented could not be in equilibrium. Just as a pencil on its tip falls one way or another, Einstein's weird universe had to expand or to contract. Wanting a static solution at all costs, Einstein introduced a locking term, Λ, in the equations that described the structure and energy content of the universe. For Einstein, the universe had to be in a steady state, with the present moment being an instant between an infinite past and an infinite future. Like several of his contemporaries, Einstein had remained Newtonian in his thinking. However, two "young wolves" working independently with Einstein's equations soon developed different views and pulled out an unexpected but amazing result.

Russian physicist and mathematician Alexander Alexandrovitch Friedmann (1888–1925) was a brilliant student of general relativity. He accepted the static solution, but explored the dynamic solutions of the equations that Einstein had ignored for philosophical reasons. Friedmann's main work was published in German in 1922 and 1924. It showed that Einstein's field equations allowed a dynamic solution for the structure of space. He did not try to connect his findings with astronomical observations of the apparent recession of the bulk of galaxies that Slipher had revealed a few years earlier. Instead, he proposed a dynamic universe where "the world's radius of curvature ... is constantly increasing in time; cases are also possible when the radius of curvature changes periodically: the Universe contracts into a point (into nothing) and then again increases its radius from a point up to a certain value, then again, diminishing its radius of curvature, transforms itself into a point, etc." To illustrate his unorthodox viewpoint, Friedmann reminded his readers of the Hindu mythology of cyclic universes and of "the creation of the world from nothing."[40] Unfortunately, Friedmann passed away, victim of the typhoid fever, when only 37 years of age. But his rather fantastic view of a swinging universe was not lost for long.

---

[39] J. A. Wheeler (with Kenneth Ford), *Geons, Black Holes and Quantum Foam*, New York: W. W. Norton & Company, 1998, p. 235.
[40] Cited in E. A. Tropp, V. Ya. Frenkel and A. D. Chernin, *Alexander A. Friedmann: The Man who Made the Universe Expand*, Cambridge: Cambridge University Press, 1993, p. 157.

Fig. 5.9 Georges Lemaître. Credit: Archives Georges Lemaître, Université catholique de Louvain, Louvain-la-Neuve, Belgium.

A young Belgian cosmologist, unknowingly and independently of Friedmann, was conducting his own exploration of general relativity. He also had something phenomenal to tell about the structure and origin of the universe. Independently of Friedmann, Georges Lemaître (1894–1966) found the expanding or contracting universe solutions to the equation of general relativity (Fig. 5.9). In 1927, that is two years before Hubble's first announcement of the velocity–distance relation, the young Belgian priest published his work in French. Lemaître had given the following (translated) title to his seminal paper: "A homogeneous universe of constant mass and increasing radius accounting for the radial velocity of extragalactic nebulae."[41] The last words clearly indicated that Lemaître was fully aware of the observational results on the recession of "nebulae." Using the data set on the velocities of galaxies measured by Slipher, he concluded that the universe was expanding, not contracting. Lemaître's work was at first ignored, until the British astronomer and physicist Arthur Stanley Eddington (1882–1944) had Lemaître's paper translated and published in the British journal, *Monthly Notices of the Royal Astronomical Society*, in

[41] Lemaître's paper appeared in French in the *Annals of the Scientific Society of Brussels*, 1927, Vol. 47, pp. 49–59.

1931.[42] This time Lemaître's work aroused wide interest. In his groundbreaking paper, the young cosmologist had derived an approximation of what was to be called later the "Hubble law" (Figs. 5.10 and 5.11). As commented by the science historian Helge Kragh: "The famous Hubble law is clearly in Lemaître's paper. It could as well have been named Lemaître's law."[43] It can firmly be established that Georges Lemaître figured out the expansion of the universe before anyone else.

At the Mount Wilson Observatory, Hubble and Humason had initiated a massive program to photograph galaxies and measure their apparent velocities. They published results confirming the expansion of the universe in 1931 (Fig. 5.10). Following the formative paper, the study of the expansion of the universe became a major field of investigation for the rest of the twentieth century and beyond.[44] Again with a groundbreaking result, the recurring hesitancy on the part of Hubble came into play: "The interpretation of red-shifts as actual velocities, however, does not command the same confidence, and the term 'velocity' will be used for the present in the sense of 'apparent' velocity, without prejudice as to its ultimate significance."[45]

The historians Harry Nussbaumer and Lydia Bieri have reviewed the papers and notes of Hubble and of Hubble's colleagues. Their conclusion is inescapable: Hubble did not believe that the high receding velocities were real, or meant that spacetime was expanding.[46] Hubble, whom many quote incorrectly as being the discoverer of the expanding universe, did not believe in this expansion. Allan Sandage wrote in 2009: "The irony, of course, is that although the discovery of the expansion is often attributed to Hubble with his 1929 paper, he never believed in its reality."[47] In the end, it is fair to say that the discovery of the expanding universe was the collective work of several men: Einstein, Friedmann and Lemaître on the theoretical front; Slipher, Humason and Hubble (unwillingly) on the observational side, with some obvious reluctance on the part of some of the players.

Expansion meant that, in the past, the universe was smaller and in a distant past, very, very small. Following on Friedmann's idea of creation out of nothing, the young and bold Lemaître invoked a new quantum theory to propose a physical and rigorous explanation for an early, very hot universe, which he called the "primeval atom."[48] It was 1931, and Lemaître's daring proposal was the birth of the Big Bang theory. Lemaître's extraordinary intuition was the existence of a primeval atom, the initial phase of the universe when

---

[42] On the controversy around the translation, see Mario Livio, Mystery of the Missing Text Solved, *Nature*, 2011, Vol. 479, pp. 171–173.

[43] H. S. Kragh, *Cosmology and Controversy, The Historical Development of Two Theories of the Universe*, Princeton: Princeton University Press, 1996, pp. 29–30.

[44] E. P. Hubble and M. L. Humason, The Velocity–Distance Relation Among Extra-Galactic Nebulae, *The Astrophysical Journal*, 1931, Vol. 74, pp. 43–80.

[45] E. Hubble and M. L. Humason, The Velocity–Distance Relation Among Extra-Galactic Nebulae, *The Astrophysical Journal*, 1931, pp. 76–77.

[46] H. Nussbaumer and L. Bieri, *Discovering the Expanding Universe*, Cambridge: Cambridge University Press, 2009, pp. 119–120.

[47] A. Sandage, in H. Nussbaumer and L. Bieri, *Discovering the Expanding Universe*, Cambridge: Cambridge University Press, 2009, p. xv.

[48] G. Lemaître, The Beginning of the World from the Point of View of Quantum Theory, *Nature*, 1931, Vol. 127, p. 706. Lemaître's pioneering paper made less than half a page in *Nature*; it is packed with insights. Lemaître's paper shared the journal page with another short article, "Insects remains in the guts of a cobra."

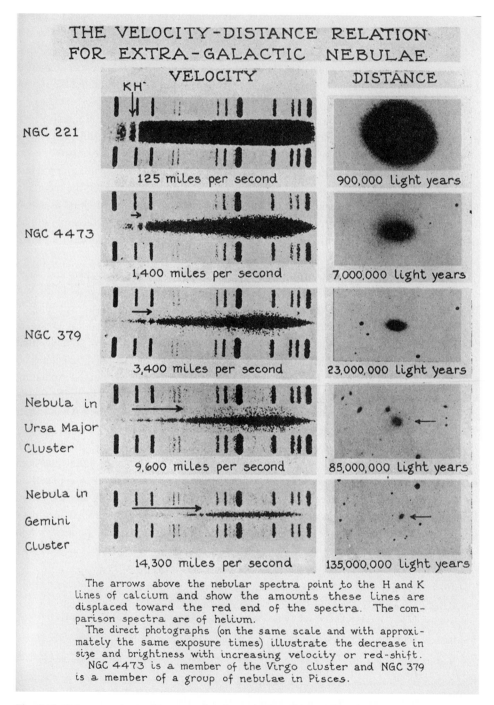

Fig. 5.10 Galaxy spectra by Humason showing galaxies of increasing distances and their spectra displaying greater receding velocities. From Humason (1936), *The Astrophysical Journal.* © AAS. Reproduced with permission.

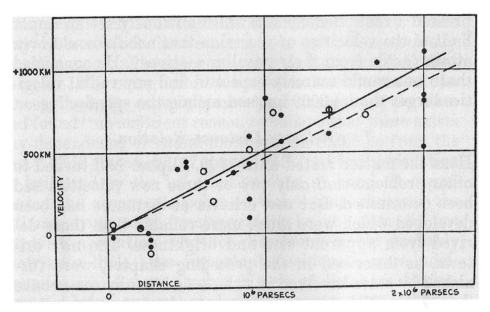

Fig. 5.11 The velocity–distance linear relation as originally presented by Hubble. From Hubble (1929), *Proceedings of the National Academy of Sciences.*

density and temperature were awesomely colossal, and the fabric of spacetime suddenly unfolded. Although Fred Hoyle described this beginning as a 'big bang' to ridicule the idea, Lemaître's revolutionary concept has been vindicated multiple times by observations.

The aftermath of the discovery of the world of galaxies and of the expansion of the universe did not end here. While trying to establish the mean density or the mass of the universe by mapping the expansion velocity of the universe as a function of time, and expecting to find a weak deceleration due to gravity, groups of researchers found that the expansion was instead accelerating. These findings, published in 1998, were confirmed by detailed spectroscopy of distant supernovae of type Ia. Such supernovae result from material that is "dumped" by a red giant star on its companion, a degenerate star (white dwarf), triggering the nuclear detonation of the star. Researchers now associate the acceleration with the 'infamous' Λ of Einstein. This time, instead of locking the universe into a static state, the new Λ fosters the physics for the acceleration under a term called "dark energy." We don't yet know what this is.

A Rip-Roaring Century

"... It seems clear that the discovery of new classes of astronomical objects is an extended process consisting of detection, interpretation, and understanding, all of which may be preceded by a pre-discovery phase."[49] Images of "nebulae" and of galaxies as they developed over centuries (Chapters 1 to 5) were key tools for the many conceptual transformations and paradigm shifts the astronomers went through.

In the last 100 years, astronomers exploring and mapping the world of galaxies have inflated the volume of the observable universe by 10^{15}, or one million billion times, and have unveiled a new form of mysterious energy: dark energy. Invoking Stigler's law of eponymy, it is safe to say that many people "discovered" galaxies; several people established the distances to galaxies; a few people found that the universe was expanding and accelerating. Among these, some individuals stand out for their outstanding clarity, audacious insights and stunning observational findings. Their main tool was astronomical images of increasing depth and precision, including amazing images of spectra of galaxies. The next three chapters will explore how images and spectra unveiled galaxies.

---

[49] S. J. Dick, *Discovery and Classification in Astronomy, Controversy and Consensus*, Cambridge: Cambridge University Press, 2013, p. 189.

6

Galaxies in Focus

Galaxies are the largest single aggregates of stars in the universe. They are to astronomy what atoms are to physics.

Allan Sandage[1]

The absorption is effective in all galactic longitudes but seems to take place mainly in a thin layer extending along the galactic plane.

Robert J. Trumpler[2]

How were Key Properties of Galaxies Discovered?

German astronomer Wilhelm Heinrich Walter Baade (1893–1960) was a visitor to the Mount Wilson Observatory from 1926 (Fig. 6.1), and he moved to work there from 1931 to 1959. Allan Sandage and Halton Arp were his two best known doctoral students, and their roles and influences will be discussed later. Having neglected to refresh the citizenship papers he had lost before World War II, he was declared an "enemy alien" of the United States at the start of World War II. During the war, Baade was restricted to the Mount Wilson and Pasadena area in southern California, but was allowed to continue working. He used the 100-inch telescope under unusual night-sky conditions, as Los Angeles and the neighboring towns were under a strict blackout.

Baade was a superb and talented observer, outstanding at getting the best photographic material, an ability he had already demonstrated when observing at the Hamburg Zeiss 40-inch reflector back in Germany.[3] His work was always of exquisite quality. Halton Arp gave a colourful description of how meticulous Baade was with his photographic work: "He insisted on having separate dark rooms in the telescopes [sic]. One for people who did direct photography like himself and the other for what he called the pigs, the spectroscopists, who slopped the dark room up and made messes and shouldn't be allowed in a top flight dark room."[4]

---

[1] A. R. Sandage, *The Hubble Atlas of Galaxies*, Washington: Carnegie Institution of Washington, 1962, p. viii.
[2] R. J. Trumpler, Preliminary Results on the Distances, Dimensions and Space Distribution of Open Star Clusters, *Lick Observatory Bulletin*, 1930, No. 42, p. 188.
[3] D. E. Osterbrock, *Walter Baade, A Life in Astrophysics*, Princeton: Princeton University Press, 2011.
[4] H. C. Arp, in an American Institute of Physics oral interview by Paul Wright, 29 July, 1975.

Fig. 6.1 Walter Baade in 1926. Credit: University of Chicago Photographic Archive, [apf6–01309], Special Collections Research Center, University of Chicago Library.

Resolving Stars in Galaxies

Using his wartime isolation "privilege," Baade pushed the depth and resolving power of photography on the 100-inch as no one else before. "There is a very considerable, broad, pretty faint, small nebulae near it; my Sister discovered it August 27, 1783, with a New-tonian 2-feet sweeper."[5] This is how William Herschel described NGC 205, one of the small elliptical galaxy satellites of Messier 31. Almost 160 years later in 1944, Baade managed to resolve the stars in the central region of Messier 31, the Andromeda Galaxy, and also in its elliptical companions Messier 32 and NGC 205.[6] Despite numerous claims by past observers, this was the first time individual stars could be seen in a galaxy. Allan Sandage included a Palomar 200-inch image of NGC 205 obtained by Baade in *The Hubble Atlas of Galaxies* to illustrate the elliptical morphological type of galaxies. In a side

---

[5] W. Herschel, On the Construction of Heavens, *Philosophical Transactions of the Royal Society of London*, 1785, Vol. 75, pp. 213–266.
[6] W. Baade, The Resolution of Messier 32, NGC 205, and the Central Region of the Andromeda Nebula, *The Astrophysical Journal*, 1944, Vol. 100, pp. 137–146.

note to the photograph, Sandage wrote "... resolution does not occur until a critical exposure time is reached, at which time the entire smooth image of the galaxy breaks up into individual stars."[7]

In a companion paper of the same issue of the journal, Baade presented stunning images of NGC 147 and NGC 185 as new members of the Local Group of galaxies.[8] He made history with his exquisite photographs of the galaxy. Baade's photographs of these nearby galaxies are some of the most important galaxy images of twentieth-century astronomy. Being able to resolve individual stars, Baade could determine their luminosities and colours. He clearly distinguished two families of stars, Populations I and II, shortened as Pop I and Pop II, as they came to be called. For the first time, a case was made for true cosmic evolution, "a process that results from the change in the 'genetic' material of successive generations of the 'species,' not just the aging of individual members of the species."[9] In this case, the species are the different stellar populations that Baade had managed to separate very clearly: (i) yellowish and reddish stars of low luminosity and a low content of chemical elements heavier than helium (i.e. "metals"), which were dominant in the central bulge; and (ii) relatively bluish stars of higher luminosity and higher metal content, which were located in the extended galaxy disk and spiral arms. In his paper, Baade commented on the resolution: "The plate reveals incipient resolution of NGC 205 quite unmistakably; but the prevailing pattern is still very soft, and the smallest elements are not yet stars but small-scale fluctuations in the stellar distribution. The resulting impression is very irritating to the eye."[10]

The images Baade had obtained were densely crowded with myriad stars. It was difficult to reproduce the details with ordinary halftone illustrations. "Baade therefore requested that actual photographic prints of NGC 185 be used in his paper in *The Astrophysical Journal*, and the editor agreed to this procedure...."[11] This was certainly an extremely rare case of an actual photograph being bound with the text of a printed article (Fig. 6.2). An unusual editorial note introduced the special image: "The photographic reproduction on the opposite page is from a negative of NGC 185 by Dr. Baade. North is toward the binding; west is at top of page. The prints were produced at Yerkes Observatory by the Misses Maude Laidlaw and Doris Blakeley, under the supervision of Dr. W. W. Morgan from a duplicate negative prepared at Pasadena by Mr. E. R. Hoge. – Editor."[12] Baade had engaged the best professionals to prepare the most appropriate production of his unique image.

---

[7] A. R. Sandage, *The Hubble Atlas of Galaxies*, Washington: Carnegie Institution of Washington, 1961, p. 3.

[8] W. Baade, NGC 147 and NGC 185, Two New Members of the Local Group of Galaxies, *The Astrophysical Journal*, 1944, Vol. 100, pp. 147–150.

[9] D. Mihalas, Baade's Resolution of M32, NGC 205, and M31, *The Astrophysical Journal: American Astronomical Society Centennial Issue*, 1999, Vol. 525, Number IC, Part 3, p. 360.

[10] W. Baade, NGC 147 and NGC 185, Two New Members of the Local Group of Galaxies, *The Astrophysical Journal*, 1944, Vol. 100, p. 140.

[11] O. Struve and V. Zebergs, *Astronomy of the 20th Century*, New York: The Macmillan Company, 1962, p. 450.

[12] W. Baade, NGC 147 and NGC 185, Two New Members of the Local Group of Galaxies, *The Astrophysical Journal*, 1944, Vol. 100, p. 148.

PLATE VI

N

W

(a)   (b)

Fig. 6.2 **Transformational Image: Baade's Photograph Resolving for the First Time Individual Stars in Nearby Galaxies.** (a) NGC 185, "best described as a slightly elongated, giant globular cluster," photographed in red light by Walter Baade using the Mt Wilson 100-inch telescope. From a photographic insert in Baade (1944b), *The Astrophysical Journal*. © AAS. Reproduced with permission. (b) NGC 185, in a fine halftone reproduction, resolved into stars and showing patches of obscuring material. From Baade (1944b), *The Astrophysical Journal*. © AAS. Reproduced with permission.

Walter Baade was not only a master at producing exquisite images, he also had a profound influence on our understanding of galaxies. And as an advocate for fine astronomical imaging, he went on to promote a new type of imaging telescope. Enjoying a deep friendship with the Estonian optician Bernhard Schmidt (1879–1935), Baade discussed new telescopic designs for improved astrophotography. He was looking for optical systems with an enlarged field of view without off-axis aberrations of coma (Chapter 3, Part B). Astonishingly, Schmidt met the challenge with a revolutionary optical design. At Baade's urging, a 18-inch Schmidt optical system was built in Pasadena and installed at Mount Palomar in 1936 where Fritz Zwicky put it to work immediately (Fig. 3.11). Thereafter, several Schmidt-type telescopes were built at observatories around the world. Alongside these significant advances in telescope optics were developments in the cameras installed at telescopic foci to detect light together with light-sensitive materials that record images. The scientific requirements and evolving technologies driving these forward often acted in synergy.

Dewdrops in the Cosmic Web

Astronomical imaging has undergone several transformations, as described in the previous chapters. Humans have used photosensitive materials for almost 200 hundred years. In the first half of the nineteenth century, we learned to impress the material and to reveal imprinted images through chemical processing. With the discovery of the electron and an understanding of its behavior, e.g. the photoelectric effect, cathode ray tubes (ancestors of the television) and image intensifiers of various sorts were put into use during the 1950s and 1960s. For example, in the early 1960s, American physicist Robert Leighton (1919–1997) experimented with the vidicon, a sort of digital television tube, to be used on spacecraft (see Fig. 8.1). Although more sensitive, vidicons never provided the field of view of the photographic plate.

By the late 1970s, semiconductors of all sorts were being designed and assembled. A profound change took place with the invention and fabrication of solid-state photosensitive devices, in particular charge-coupled devices (CCDs). A CCD is a solid-state device that converts light into an electrical signal with a very high efficiency (close to 100%). The invention triggered the era of digital electronic imaging. During the 1980s, CCDs were rapidly adopted and improved by astronomers in their drive for the higher sensitivity necessary for detecting low fluxes of photons. Nevertheless, much had been learned from using a good old photographic emulsion (Chapter 10), and it needs to be said that photographic plates had one huge advantage over CCDs: they were largely free of artefacts and had good cosmetics. However CCD have an enormous advantage: they are able to record photons ten times more efficiently than the most sensitive photographic emulsions. In the wave of technological spinoffs of the past two decades, CCDs have become ubiquitous in our daily life; they are part of the minuscule cameras of cell phones. In astronomy, CCDs have been hugely transformative and have opened up remarkable imaging opportunities (see Fig. 3.12 and Fig. 3.13).

Whether assembled in giant clusters or lined up in huge filamentary structures in an enormous spider's-web pattern, galaxies weave the fabric of space in the universe. Each galaxy is a complex assembly of matter. Individual galaxies are giant stellar, gaseous and dust systems held together by the force of gravity. The gravitational potential that holds galaxies and clusters of galaxies together is shaped by something subtle whose nature remains unknown. The descriptor we have for this "thing" is vague: it is called dark matter, or better still, invisible mass. Dark matter anchors everything we see and it seems to be ubiquitous, and dominant in large scales (Chapter 8). Although luminous or absorbing matter in galaxies represents only a fraction of the total mass, we have learned from it almost all we know about galaxies, and several properties of dark matter itself.

Fulfilling a century-old and once hopeless quest, twentieth-century large-aperture telescopes, operating under good seeing conditions, finally resolved individual stars into galaxies (Fig. 6.2). By studying the images of galaxies obtained in both visible and infrared light, astronomers have been able to infer many important properties of galaxies and elucidate large-scale properties of the universe, such as those reviewed below.

Imaging Darkness and Magnetic Fields

The gaseous state is one of the forms of interstellar matter. The other component of galaxies, hinted at in the previous sections, is dust, which is permeating interstellar space in the form of tiny particles of micron size or less, made of icy or mineral solids. Dust makes up about 1% of the mass of the gas component. Although this sounds unimportant, dust holds several keys for chemical processing of elements and molecule formation. It also affects the way we view galaxies.

The prevailing presence of dust was first inferred from its obscuring properties (Chapters 3 and 4). Dispersed between stars, interstellar dust dims the light of distant stars; these appear fainter than they would be with no intervening material. Particularly noticeable are the dense dust clouds blocking starlight and producing dark nebulae. In the early twentieth century, American astronomer and astrophotographer Edward Emerson Barnard (1857–1923) highlighted the dramatic masking effect in his magnificent book, *A Photographic Atlas of Selected Regions of the Milky Way*.[13] Like others, Barnard initially thought that the dark spots in the sky were holes in space, places with no stars.[14]

It took some time for some astronomers to realize that the dark patches seen on photographic plates were real and not photographic defects. The Italian astronomer Angelo Secchi (1818–78) talked about "canals," while the Dutch amateur astronomer Cornelis Easton (1864–1929) used more visual depictions, "great rift," "chasms" or "dark streams."[15] Despite strong evidence, astronomers of the day did not believe in the existence of interstellar obscuration.[16] "Throughout the 1920s there was general disdain for the whole idea of an absorbing medium. It is one of the most astonishing examples of wishful thinking in the history of astronomy."[17] However, Heber Curtis and others agreed that certain spiral nebulae, seen edgewise showing a dark lane running down the length of the spirals, could be explained as due to a band of absorbing or occulting matter, and similarly for the dark patches masking the middle portions of the Milky Way (see Plate 6.1).[18] It was Max Wolf who, in 1923, showed this obscuration to be caused by dust and not by lack of stars.[19]

In a brilliant 1930 paper, Swiss–American astronomer Robert Julius Trumpler (1886–1956) demonstrated that dust of micron size also produces an artificial change of the colour of starlight we receive.[20] This effect was called "reddening," because dust absorbs more

[13] E. E. Barnard, *A Photographic Atlas of Selected Regions of the Milky Way*, Washington: Carnegie Institution of Washington, 1927.

[14] E. E. Barnard, On the Dark Markings of the Sky with a Catalogue of 182 Such Objects, *The Astrophysical Journal*, 1919, Vol. 49, pp. 1–23.

[15] C. Easton, A Photographic Chart of the Milky Way and the Spiral Theory of the Galactic System, *The Astronomical Journal*, Vol. 37, p. 109–110.

[16] I. Roberts had noticed and commented on the "broad dark band" that shuts out the light of the central condensation. See images of NGC 3628, 4565 and 4594 on Plate XX in *Photographs of Stars, Star Clusters and Nebulae*, London: Knowledge Office, 1899, Vol. II.

[17] J. D. Fernie, The Historical Quest for the Nature of the Spiral Nebulae, *Publications of the Astronomical Society of the Pacific*, 1970, Vol. 82, pp. 1209–1210.

[18] H. D. Curtis, A Study of Occulting Matter in the Spiral Nebulae, *Publications of the Lick Observatory*, 1918, Vol. 13, Part II, pp. 45–54, and seven plates.

[19] M. Wolf, Über den dunklen Nebel NGC 6960, *Astronomische Nachrichten*, 1923, Vol. 219, pp. 109–116.

[20] R. J. Trumpler, Absorption of Light in the Galactic System, *Publications of the Astronomical Society of the Pacific*, 1930, Vol. 42, pp. 214–227.

light of shorter than of longer wavelength. Interstellar reddening is similar to the effect of the Sun or the Moon appearing redder before setting or after rising, because of dust in the lower layers of our atmosphere scattering the blue light.

The magnificent luminous band crossing the night sky has filled our ancestors with wonder since prehistoric times. The obscuring dust can easily be seen when viewing the Milky Way from a dark site, especially from the southern hemisphere where the center of the Milky Way crosses close to the zenith. Extended patches can be seen where there appears to be fewer stars cutting across large sections of the Milky Way (Plate 6.1). Australian aborigines interconnected these dark patches and imagined a giant celestial emu stretching along the celestial vault. When observed in the more penetrating near-infrared light instead of visible light, these patches become transparent: a multitude of embedded stars, and many beyond the absorbing cloud, are revealed.[21] Today, the dusty plane of the Milky Way is related to the "zone of avoidance" where galaxies appear absent, an apparent distribution that deeply puzzled observers of the nineteenth century (Chapter 5).

Starlight heats interstellar dust. The tiny grains settle to a temperature of a few hundred kelvins, the equilibrium value depending on the composition of the grain and its energy loss rate. Hence, heated dust radiates and produces a thermal emission in the infrared domain. Dust clouds, obscuring at optical wavelength, become luminous at longer wavelengths, and are easily observable in the mid-infrared at wavelengths of 5 microns and more (Fig. 0.4). This happens for dust near massive hot stars or close to the central active galaxy nuclei.

Although dust represents only a small proportion of the mass of the galaxies, these tiny solids hold an unusual importance. Their affinity for certain elements is strong enough that the grains just suck up the available heavy elements. This chemical capture locks a significant fraction of the heavy elements dispersed in the interstellar medium, depleting the interstellar gas of elements such as deuterium, silicon and iron, which enter into the composition of the interstellar dust grains. While dust traps metals, it betrays a fundamental universal force permeating interstellar space. The small grain can become magnetized.

A large-scale and low-intensity magnetic field pervades the interstellar space of galaxies. Its strength is weak, a million times weaker than Earth's magnetic field, but it spreads on a scale of light-years. As clouds of interstellar gas and dust move around, they collide with each other. Like invisible springs, the embedded magnetic fields become compressed. Expanding bubbles, driven by the stellar winds and supernovae explosions of massive stars, also compress the gas and dust and its magnetic field. As this happens, electrically charged particles, trapped in the magnetic field lines, are accelerated to higher energies, like pebbles spun with a slingshot. It was the Italian physicist Enrico Fermi (1901–1954) who showed that moving magnetic fields have the net effect of accelerating charged particles to extremely high energies.[22] When the accelerated particles reach high enough energies, they escape their cloud and propagate through space as "cosmic rays." These are electrically

---

[21] The near-infrared domain corresponds to wavelengths of 1 to 5 microns, and the mid-infrared to wavelengths of 5 to 25 microns.
[22] E. Fermi, Galactic Magnetic Fields and the Origin of Cosmic Radiation, *The Astrophysical Journal*, 1954, Vol. 119, pp. 1–6.

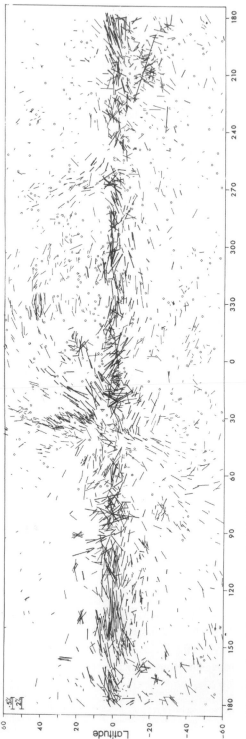

Fig. 6.3 Map of starlight polarization for 7,000 stars across the sky. The short lines indicate the strength and direction of the polarization (E-vector), indicative of the projected magnetic field permeating the interstellar medium. The galactic latitude is shown as the y-axis, and galactic longitude as the x-axis, with 0° corresponding to the direction of the galactic center. From Mathewson and Ford (1970), *Memoirs of the Royal Astronomical Society*.[23]

[23] D. S. Mathewson and V. L. Ford, Polarization Observations of 1800 stars, *Memoirs of the Royal Astronomical Society*, 1970, Vol. 74, pp. 139–182.

charged particles, mainly protons and electrons, microscopic bullets traveling at velocities close to the speed of light. They are a natural source of radioactivity, and when they collide with living cells, they can trigger mutations. Hence cosmic rays may have played a significant role in the evolution of living species. Details of how energetic particles also produce electromagnetic waves in the radio domain of the electromagnetic spectrum are given in Chapter 7.

A special imaging technique, polarimetry, gives us the means to measure and map the magnetic fields. Here is how it works. The small dust grains are weakly magnetized. As tiny magnets, they line up with the general magnetic field, like the needle of the compass lines up with the Earth's magnetic field. Light going through or scattered by the dust varies in intensity with orientation as viewed against the plane of the sky: light is polarized, i.e. it is more intense at a certain angle as it passes more easily through dust grains aligned in a given direction. The degree of polarization is a measure of strength of the magnetic field that lines up the grains. By analyzing the polarization of starlight over many directions in the sky, a map of the Milky Way's magnetic field can be made, an informative non-homomorphic representation (Fig. 6.3; see also Plate 7.2).

Furthermore, an active surface chemistry makes the interstellar dust grains microfactories of complex molecules. The interstellar molecular products go from rather simple radicals, such as OH or CH^+, or molecular hydrogen, to a whole range of molecules like carbon monoxide, water or more complex molecules made up of as many as 17 atoms, and even amino acids. The interstellar medium is particularly efficient at making water ice. For example, water on Earth was produced in interstellar clouds prior to the formation of our solar system. Molecules emit mostly in the infrared and radio domain of the electromagnetic spectrum, and astronomers are able to make images of molecular clouds at those wavelengths.

Shaping the Milky Way

If the nearest star is the Sun, the nearest galaxy is the Milky Way, and we are embedded in it. In one of the great breakthroughs of early twentieth-century astronomy, the American astronomer Harlow Shapley proved, in 1918, that we were not at the center of this giant system of stars as many believed until then.[24] Shapley assumed zero dust and perfect transparency. He also mistook the short-period variable stars, which he used for determining the distance, for brighter ones, and hence grossly overestimated the size of our Milky Way by a factor of three. However, his basic approach and his conclusion were correct. He derived that our Sun and solar system are in orbital motion around the galactic center that lies in the direction of the constellation of Sagittarius. It is now well established that the Sun and its planetary system is located at about 27,000 light-years from the center of the

---

[24] H. Shapley, Studies Based on the Colors and Magnitudes in Stellar Clusters – Seventh Paper: The Distances, Distribution in Space and Dimensions of 69 Globular Clusters, *The Astrophysical Journal*, 1918, Vol. 48, pp. 154–181.

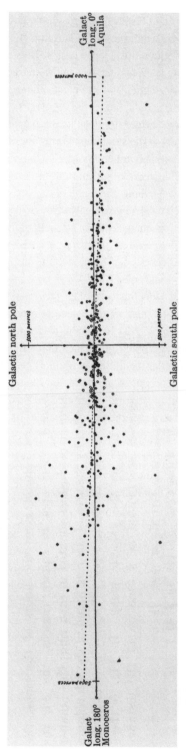

Fig. 6.4 Projection of 334 open star clusters (within 1,000 parsecs of the Sun) projected on a plane perpendicular to the galactic plane. The dotted line marks the plane of symmetry of the open clusters. From Trumpler (1930), *Lick Observatory Bulletins.*

Milky Way. At the average speed of 828,000 km/h (230 km/s), it takes us about 225 to 250 million terrestrial years to complete a full galactic revolution.

Knowing our Milky Way and its constituents has been an essential step in understanding other galaxies, including those of different histories, shapes and masses. But the epistemic process also works in reverse. It became easier to chart our own galaxy and describe its shape once the extragalactic nature of "nebulae" was established. The Lick Observatory astronomer Robert Trumpler found inspiration in images of other spirals. He figured out the overall shape of the Milky Way system by mapping the distribution of star clusters. In his second seminal paper, also published in 1930, Trumpler presented the results of a study of 334 open star clusters, groups of stars that formed coevally from the same molecular cloud. By mapping the distribution of these clusters, Trumpler showed that the Milky Way had a flattened disk shape (Fig. 6.4). "The hypothesis supports the view that our Milky Way system is a highly resolved spiral nebula, a right-handed spiral as seen from the galactic north pole, of dimensions similar to those of the Andromeda nebula."[25]

Crystallizing Galaxy Shapes

As has been shown in the previous chapters, eye observations, drawings and photographs helped to find order in the span of galaxy shapes. In his 1811 article, William Herschel drew forms of "nebulae" where basic galaxy silhouettes are recognizable (Chapters 1 and 2). Half a century later, the Birr Castle observers sketched most of the key shapes that twentieth-century observers used in designing classification schemes (Chapter 9). With the ability to measure distances and the understanding of how dust affects our viewing, exploring the shapes of the various galaxies became an important area of research. Ellipticals and disks were the two main categories of galaxies quickly identified, both presenting varying degrees of flattening for ellipticals and central concentration for spirals or disk systems.

Contrary to Curtis' early insight, there were more than just spirals in the extragalactic world. As discussed briefly above, the contents and shapes of galaxies are determined by nature, inherited conditions at birth, and by nurture, later events from interaction with their environment. Galaxies can be rich or poor in interstellar gas. Their present gaseous and dust content reflects the conditions that have prevailed throughout galactic history. These are driven, first, by initial states such as the mass and angular momentum of the proto-galaxy, and second, by the environment, factors such as intergalactic density, or interaction with neighbors at birth or later.

The New Zealander–American astronomer Beatrice Tinsley (1941–1981) produced fundamental work that linked stellar evolution, gas consumption and the integrated properties of galaxies as they evolved. She found that "while these calculations cannot prove that the sequence irregular–spiral–elliptical is not an evolutionary order, they show that all the principal galactic types may have originated at the same time, but with some differences in

---

[25] R. J. Trumpler, Preliminary Results on the Distances, Dimensions and Space Distribution of Open Star Clusters, *Lick Observatory Bulletin*, 1930, No. 42, pp. 154–188.

physical conditions that led to different stellar birth rates."[26] Modern astrophysicists have been able to disentangle these various factors using a range of techniques. Images of galaxies have helped provide insight into the process of proto-galaxies transforming into the galaxies seen today.

Galaxies as Interstellar Gas Processors

Although their morphology and mass may have changed considerably over aeons, most galaxies appear to have been in place for billions of years. In addition to billions, and sometimes trillions of stars, galaxies hold variable amounts of gas and dust, current contents being mostly dictated by how galaxies initially came together. Primordial collapse of giant clouds took place at rates that depended on the large-scale properties of the primordial units and on the environment. These materials have been recycled in and out of stars to form successive generations of stars. This lifecycle rhythm has determined the morphological properties of the galaxies we observe today, as inferred by Tinsley.

The process continues today at various rates. With each generation of stars, the heavy-element content of stars and gas clouds is enhanced. Adding to this "closed-box" processing, the merging of galaxies, cannibalizing each other, has affected their evolution to a degree that we are still trying to understand. In today's galaxies, a large fraction of the primordial gas has been used up, having condensed into stars and planets, or lost to intergalactic space. As this process has happened at different rates, interstellar gas and dust represent a varying fraction of the visible mass of galaxies of the present day. In some galaxies, ellipticals for example, the gas fraction is almost zero. In the flatter spirals, the collapse of the proto-galaxy was slower; the gas content is still relatively high, several percent of the total mass.

"Starburst galaxies" or galaxies currently undergoing intense star-formation episodes make for spectacular images (see Plate 6.2 and Plate 11.1). We find them at both ends of the scale of mass, among small irregulars and giant ultraluminous galaxies. They harbor a high proportion of young, massive stars that sculpt the interstellar medium of these galaxies. The irregular and filamentary appearance of these galaxies is indicative of a highly turbulent interstellar medium ploughed by shocks and expanding super-bubbles, phenomena associated with the evolution of giant clusters of massive stars.

Order from Chaos

If a proto-galaxy originated from an initially slowly rotating cloud, its collapse was fast and led to an intense firework with a huge number of stars formed in rapid sequences of a few hundred million years long: most of the initial gas condensed into stars of all masses. This scenario can be viewed as a grand scale version of what Immanuel Kant and Pierre

---

[26] B. J. Tinsley, Evolution of the Stars and Gas in Galaxies, *The Astrophysical Journal*, 1968, Vol. 151, p. 558.

Simon Laplace imagined for the formation of the solar system. Then, the large number of very massive stars evolved rapidly into supernovae and blew away the unused gas into the circum-galactic and intergalactic environment. The dried-up descendants of these rapidly forming stellar fireworks are observed today as elliptical galaxies, or elliptical as per their silhouettes. Physically, they are spheroidal or oblate systems (see Plate 6.3).

In ellipticals, stellar orbits occupy no preferential plane; they spread out in all directions and the multitude of stars move either in prograde or retrograde directions. One can imagine such systems as being somewhat analogous to gigantic swarms of bees collecting around a beehive, or for the stars, around a common center of mass.

Selected directions or planes may host a denser traffic due to dynamic resonances, or gravitational potential asymmetry induced by the capture of a smaller galaxy. Internal resonant modes determine a variety of three-dimensional shapes, from the perfect spheroidal to ellipsoidal or even prolate. The whole galaxy slowly spins on itself, especially in the central parts where rapidly rotating cores have been mapped. There are even cases of counter-rotating cores, the inner core rotating in reverse from the main body of the galaxy.

Ellipticals encompass the whole range of galaxy masses, including the most massive in the universe. The mammoth elliptical Messier 87, near the center of the Virgo cluster of galaxies, 53 million light-years away, has more than one trillion stars and is surrounded by 30,000 globular clusters (Fig. 6.5). The least massive galaxies are also ellipticals: the dwarf spheroidal Leo I galaxy, 820,000 light-years away, is a tiny member of the Local Group of galaxies. It has a few tens of millions of stars and three known globular clusters. Likewise, the Sculptor dwarf galaxy harbors only a few million stars and it can hardly be distinguished from a blown-up globular cluster.

Disks from Orderly Traffic

Proto-galaxies with significant initial angular momentum collapsed more slowly than those that produce ellipticals. This led to a slower assembly of stars, with most of the gas assembling into a flat disk. The accumulation of gas continued over a much longer period than for rapidly collapsing spheroidal systems. Despite mass loss, disk galaxies have been able to retain a generous gaseous reservoir. Hence, star formation is still continuing in gas-rich galaxies such as ours, particularly in the spiral arms, regions of higher gas density (see Plate 6.4).

Consequently disk galaxies are much flatter than ellipticals and most of their stars are revolving in the same direction around the center of mass. All spirals and many large irregulars are disk galaxies. Like pizzas, some are flatter, some bulkier.[27] One of the most striking features of disk galaxies is their spiral arms as discovered in Messier 51 by William Parsons in 1845. The spiral pattern is not due to stars or gas flowing along the arms, inward or outward. The "spirality" is a resonance pattern (Fig. 6.6). The spiral shape arises as a vibration

---

[27] In spirals, the diameter-to-thickness ratio of the disks varies from about 100 to 1 for the most flattened disks to 10 to 1 for the bulkier disks.

Fig. 6.5 Messier 87 in the Virgo cluster about 54 million light-years away. It is the most massive galaxy in the local universe. This image by the Hubble Space Telescope shows the jet of energetic charged particles coming out of the center. Credit: NASA, ESA.

mode of the whole disk system, which induces a concentration of mass at certain *loci*, which lines up; this leads to the beautiful spiral pattern. It is like a traffic jam, with a pile-up of gas at specific locations along the galactic orbit. The resonances last a few million years and are accompanied by an increase of density of the interstellar gas by a factor of at least two or three, enough to locally enhance dust and star formation. With the *loci* of overdense gas and dust, spiral arms are cradles of young stars and nebulae, which enhance their visibility. While spiral arms correspond to transient regions of stellar formation, the underlying older stellar population remains more smoothly distributed. Infrared images, which highlight the older stellar population, do not show the spiral pattern as strikingly.

Plate 1.1 Messier 31 and its companion Messier 32 and NGC 205 in Andromeda. Montage with the Moon superimposed on the same field. Credit: Adam Block/Mount Lemmon SkyCenter/University of Arizona.

Plate 4.1 NGC 2997 as a colour-coded image of relative brightness levels. Credit: ESO.

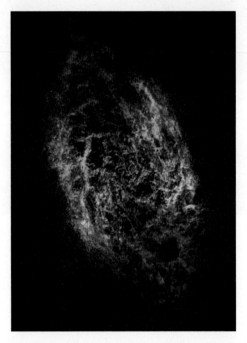

Plate 4.2 Map coded in false colours to display the velocities of the neutral hydrogen gas in the galaxy Messier 33. The blue colour represents gas approaching us and red colour, gas receding from us. Credit: National Radio Observatory/AUI.

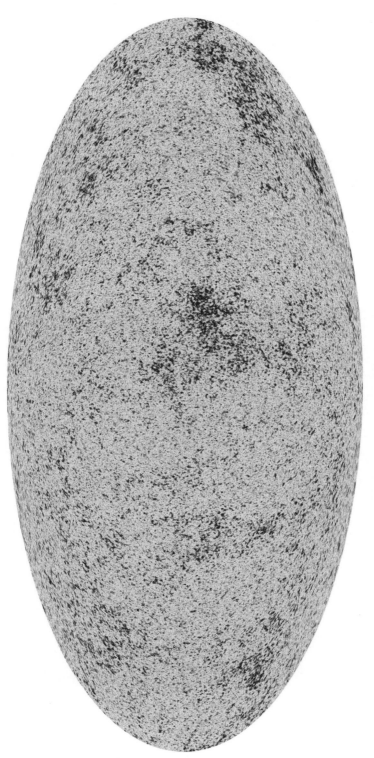

Plate 4.3 Map of the temperature fluctuations of the cosmic microwave background radiation measured at the Planck Observatory. Credit: ESA/Planck Collaboration.

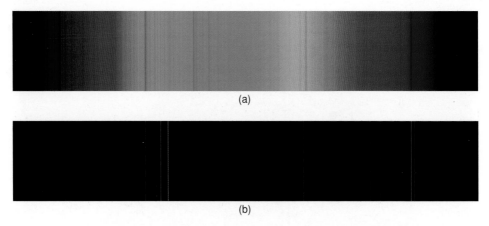

(a)

(b)

Plate 5.1 (a) Continuum spectrum showing the bright range of colours and dark absorption lines of a solar type star; (b) Spectrum of planetary nebula NGC 7027. The dominant lines are those of ionized oxygen in the green region and of hydrogen in the red. Credit: Kevin Volk, STScI.

Plate 5.2 V1 observed with the Hubble Space Telescope, ninety years after Hubble's original discovery. Credit: NASA, ESA, and the Hubble Heritage Team (STScI/AURA).

Plate 6.1 Central part of the Milky Way above ALMA antennae. The Milky Way appears like an edge-on galaxy with the striking obscuring band crossing the luminous band. Credit: ESO.

Plate 6.2 Magellanic irregular NGC 4449 about 12 million light-years away, imaged with the Hubble Space Telescope. Star-forming regions are shown in pink. Like the Large Magellanic Cloud, it has a general bar shape. Credit: NASA, ESA, Alessandra Aloisi and the Hubble Heritage Team.

Plate 6.3 Giant elliptical galaxy ESO 325-G004 at the center of the galaxy cluster Abell S07040 at 450 million light-years away. The cluster hosts a range of galaxy types. Credit: NASA, ESA and the Hubble Heritage Team (STScI/AURA); J. Blakeslee, Washington State University.

Plate 6.4 The barred spiral galaxy NGC 7424, which is of a similar morphological type as the Milky Way. Credit: Gemini Observatory.

Plate 6.5 The Large Magellanic Cloud in infrared light from dust emission observed by the Herschel Space Observatory. The image is a combination of images at wavelengths of 100, 160 and 250 microns. Credit: ESA, NASA.

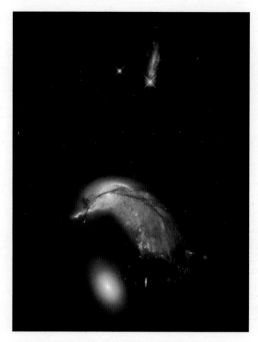

Plate 6.6 Interacting galaxy pair Arp 142 at 230 million light-years. The spiral NGC 2936 is being torn apart by the elliptical galaxy NGC 2937. Credit: NASA, ESA.

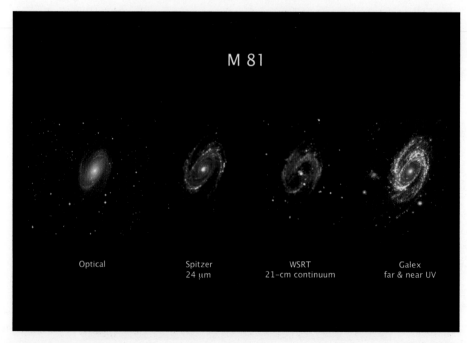

Plate 7.1 Multiwavelength images of the spiral galaxy Messier 81. From left to right: optical, mid-infrared, radio continuum and ultraviolet images. Credit: NASA, Caltech.

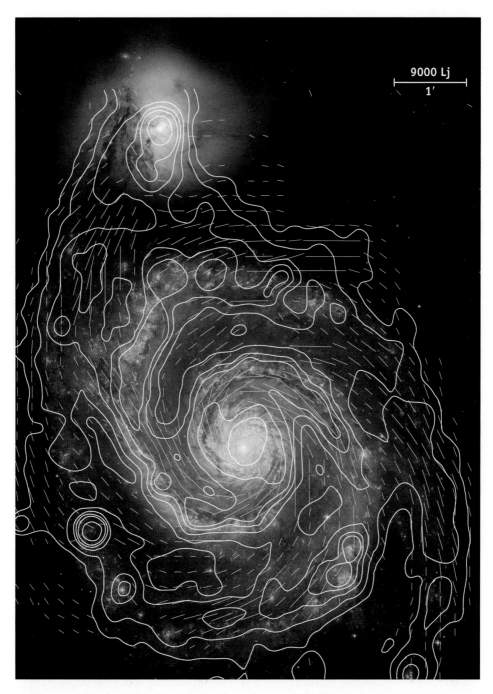

Plate 7.2 Contour map of the 6-cm synchrotron radiation observations overlaid on a Hubble Space Telescope optical image of the spiral galaxy Messier 51. The short lines (B-vectors of polarized emission) indicate the strength and direction of the magnetic field. The magnetic field is a million times weaker than that of the Earth, but is spread out on a gigantic scale. Credit: Andrew Flechter (University of Newcastle) and Rainer Beck (MPIfR Bonn), NASA/ESA. Graphics: Sterne und Weltraum.

Plate 7.3 The aperture-synthesis Very Large Array radio telescope in New Mexico. Credit: Image courtesy National Radio Observatory/AUI.

Plate 7.4 Aperture synthesis radio image (pink) of the radio galaxy Hercules A obtained with the Very Large Array. It is at a distance of 2.1 billion light-years. The superposed image in the visible region has been obtained with the Hubble Space Telescope. Credit: National Radio Astronomy Observatory, NASA, ESA and the Hubble Heritage Team.

Plate 7.5 Left, the nearby spiral galaxy NGC 2403 (8 million light-years away) observed in the 21-cm radio line of hydrogen, and on the right the galaxy as viewed in visible light. Credit: Image courtesy National Radio Observatory/AUI.

Plate 7.6 Composite image of the spiral galaxy NGC 4258 at 23 million light-years distance: X-rays from the Chandra X-ray Observatory in blue, radio emission from the VLA in purple, optical image from the Hubble Space Telescope (yellow and blue), and infrared from the Spitzer Space Observatory in red. Credit: NASA, ESA.

Plate 7.7 Composite image of the galaxy ESO 137–001 with its giant gas wake visible with X-rays. The streaming gas has been stripped from the galaxy as it moves through the intracluster hot gas. The optical image is from the Hubble Space Telescope. Credit: NASA, ESA/Chandra X-ray Observatory.

Plate 8.1 Galaxy cluster Abell 383 at 2.3 billion light-years. The optical field showing the galaxies superimposed on the diffuse X-ray-emitting gas. Credit: X-ray – NASA/CXC/Caltech/A. Newman et al./Tel Aviv/A. Morandi & M. Limousin; Optical – NASA/STScI, ESO/VLT, SDSS.

Plate 8.2 The galaxy cluster Abell 370 at about 6 billion light-years away, lensing more distant galaxies, which appear as distorted arclets. Credit: NASA, ESA/Jennifer Lotz.

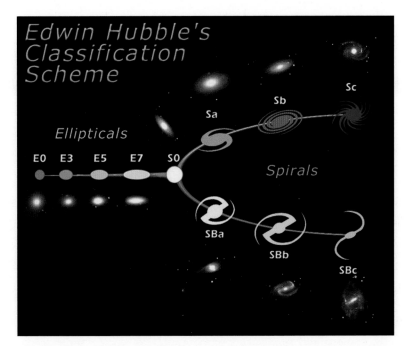

Plate 9.1 Schematics and images of galaxies illustrating the Hubble sequence, with ellipticals at the left, lenticulars (S0) at the junction, and the two branches of spirals (normal on top and barred at bottom) on the right. Irregulars are not shown. Hubble tuning-fork classification concept borrowed from John Reynolds' and James Jeans' earlier proposals. Credit: NASA, ESA.

Plate 10.1 UGC 1801. Hubble Space Telescope image of the spiral galaxy with its tidally distorted disk and its companion UGC 1813. The pair is roughly 300 million light-years away. Credit: NASA, ESA and the Hubble Heritage Team.

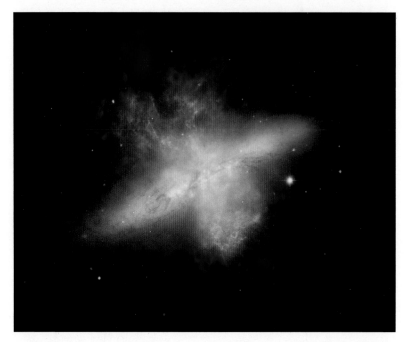

Plate 11.1 Messier 82. Composite image of the starburst galaxy in the visible (Hubble Space Telescope), infrared (Spitzer Observatory) and X-ray (Chandra X-ray Observatory) spectral regions. Credit: NASA, ESA.

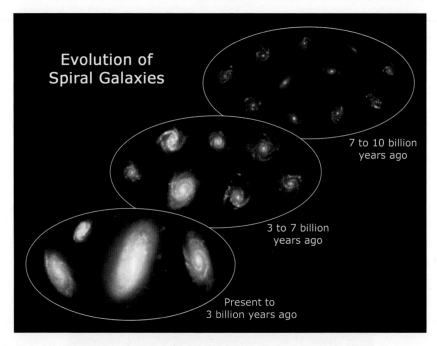

Plate 11.2 Spirals viewed back in time. Credit: NASA.

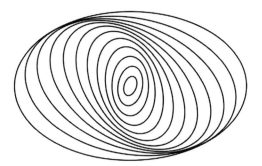

Fig. 6.6 The formation of spiral arms. Graphic illustrating how slightly elliptical orbits pile up to produce the apparent spiral pattern. Credit: Wikipedia Commons, Dbenbenn/User: Mysid.

But spirals are complex, with various components. We saw how Walter Baade could distinguish families of stars by their spatial distribution, colours and kinematics, suggesting different chemical evolution stages and ages.[28] The structural components are identified using Baade's criteria. The most extended component of a galaxy is the exponential disk, the flattened region whose brightness diminishes exponentially from the center to the edge. They often have a centrally inflated bulge where stars revolve around the center on orbits similar to those of elliptical galaxies. They also possess a halo that comprises older and high-velocity stars. The spirals with large bulges are more massive than those with small bulges. Moreover, the strength and amplitude of the spiral arms is inversely proportional to the size of the bulge, so galaxies with small bulges have the most prominent arms.

Galaxies Hit with a Jolt

Recognizing barred spirals as a distinct class of spirals, Heber Curtis introduced the subclass that he named "Φ-type spirals."[29] Indeed, in addition to their spiral pattern, disk galaxies are prone to another important instability or mode of vibration, the bar instability. The trigger for this mode has various causes, for example the presence of a companion galaxy or the passage of a nearby galaxy, which distorts their gravitational potentials. A majority of the stars in the central part of the main galaxy are slowly deviated from their circular orbits around the center and move into elongated trajectories, along near-radial orbits. Stars, gas and dust clouds then swing from one side of the disk to the other, as if carried by a huge pendulum. This is the bar (Plate 6.4).

This creates havoc in the traffic. Because interstellar clouds are relatively large, they collide, leading to the compression of gas clouds, shocks in the dust and loss of angular momentum. The gas and dust clouds fall toward the center of the galaxy where they feed a circumnuclear ring of enhanced star formation. More spectacularly, the gas is slowly

---

[28] Walter Baade, *Evolution of Stars and Galaxies*, Cecilia Payne-Gaposchkin, ed., Cambridge: Harvard University Press, 1963.

[29] H. D. Curtis, Descriptions of 762 Nebulae and Clusters Photographed with the Crossley Reflector, *Publications of the Lick Observatory*, 1918, Vol. XIII, Part I, p. 11.

sucked in by a central black hole. The visual effect of this instability can be quite striking. The collective appearance of the stars on those radial orbits is a prominent elongated feature, called a bar. This is not a solid feature but a herd of billions of stars moving collectively along weird orbits. The bar may be hidden, especially in highly tilted or dusty disks. Therefore, it is more tractable at infrared wavebands. The bar can be a conspicuous feature.

Our Milky Way is a barred galaxy probably triggered and maintained by the presence of our nearby companions the Magellanic Clouds, themselves irregular galaxies (Fig. 2.9).

Islets Adrift

If ellipticals and spirals seem to fill the universe, spread between and around them is a multitude of smaller systems. Many display amorphous shapes, as if they had been left over during the making of the larger galaxies. They are hard to sort (Chapter 9). Small galaxies, "irregulars," cannot be classified as either elliptical or spiral, and irregulars are indeed a mix bag. Best known are the small companions to the Milky Way, the Large and Small Magellanic Clouds. Interestingly, recent observations at infrared wavelengths have shown that the Large Magellanic Cloud gas and dust distribution resembles that of disk systems (see Plate 6.5).

Small galaxies can be gas-rich or almost devoid of any gas. In the latter instance, this is due to the fact that the masses of the small galaxies were probably too weak to gravitationally retain their warm or hot interstellar gas, which boiled off. Once the intense star-formation episode has waned, and the stellar orbits have stabilized after a few billion years, they became dwarf ellipticals.

Galaxies in Excited States

Finally, many galaxies do not fit the simple categories of ellipticals, disk spirals or irregulars because they are in strongly perturbed states, due to a close interaction or a recent merger (see Plate 6.6). The perturbed states involve galaxies of all classes and masses. They give rise to a wide range of shapes, some of them really weird (Chapter 10).

Deep observations by large ground-based telescopes, and in space by the Hubble Space Telescope and the Spitzer Space Telescope, reveal that "peculiar" galaxies were much more common in the early universe.[30] In these images, we see objects as they appeared 6 to 12 billion years ago when most galaxies seem to appear most peculiar. This is due both to their high rate of star formation and the more frequent merging and interaction in a smaller universe.

In the wake of Australian astronomer David Malin's pioneering work during the 1980s, new imaging techniques and processing have evolved into even more powerful discovery

---

[30] The principal large ground-based telescopes capable of reaching to the edge of the observable universe are the two 10-m Keck telescopes (Hawai'i), the two 8-m Gemini telescopes (Hawai'i and Chile), the Subaru 8-m Telescope (Hawai'i) and the four 8-m telescopes making up the Very Large Telescope (Chile).

Fig. 6.7 Elliptical galaxy NGC 474, 100 million light-years distant, shows multiple structures, likely of tidal origin. Credit: Canada-France-Hawaii Telescope Corporation.

tools. Objects that had been imaged dozens of times before turned out to be much more complex (Fig. 6.7). Faint, giant stellar rings and tidal tails were found around many normal galaxies, indicative of a convoluted dynamic history. Images have played a crucial role in viewing and understanding interactions between galaxies. Ubiquitous large-scale tidal effects had to be taken into account, and hence the role of merging and galaxy cannibalism in galaxy evolution. This opened a new research field, with astrophysicists modeling galaxies using computer simulations (Fig. 6.8).

After the first decades of the twentieth century, the "riddle of the nebulae" faded quite suddenly. Centuries-old controversies, such as the variability of "nebulae" and their resolution into stars, just evaporated as the nature of "nebulae" was finally resolved. New imaging techniques and ways of analyzing starlight, for example spectroscopy, resulted in a spectacular leap in our understanding of the world of galaxies. Powerful giant reflectors located at exquisite mountain sites and telescopes in space finally allowed the resolution of stars in the nearest galaxies. The transformations in physics pulled astronomy away from "the

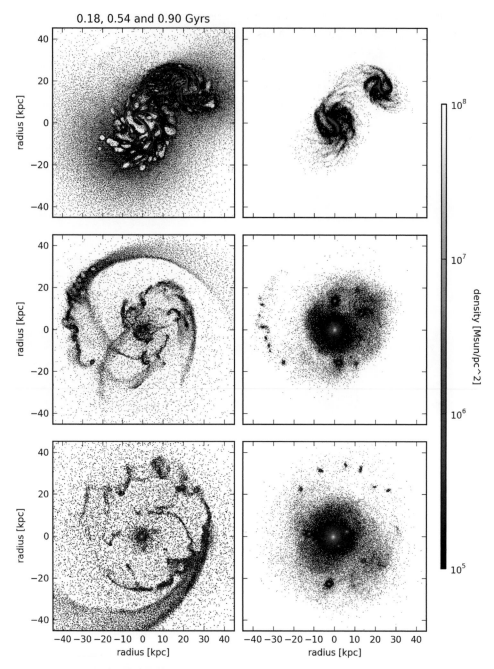

Fig. 6.8 Galaxies generated by computer simulation. Numerical simulation of the evolutionary stages (at 0.18, 0.54 and 0.90 billion years after the start of the merging) of two disk galaxies merging: left panels show the gas and right panels, the stars (compare with photograph of NGC 474, Fig. 6.7). Courtesy of Amélie Dumont and Hugo Martel.

classical, Herschelian tradition of astronomy."[31] With the discovery of the expansion of the universe, Einsteinian spacetime superseded the Newtonian cosmos. Quantum theory became the engine for the new astrophysics, furnishing precise tools to interpret the spectra of stars and galaxies. The new physics enabled technologies that opened a new window into the universe, leading to surprising new ways to look at galaxies. And gave birth to new disciplines, radio astronomy and X-ray astronomy.

[31] H. S. Kragh, *Conceptions of Cosmos, From Myths to the Accelerating Universe: A History of Cosmology*, Oxford: Oxford University Press, 2007, p. 120.

7

A Symphony of Waves

In retrospect, the serene astronomical landscape holding sway before World War II had begun to show fault lines even before the war – although few seemed to notice. The first signs came from a puzzling observation of radio noise emanating from the Milky Way's center that Karl Jansky, a radio engineer at the Bell Telephone Laboratories in Homldel, New Jersey, had discovered in 1932 . . .

Martin Harwit[1]

Using his own funds, he [Grote Reber] was able to design and construct novel apparatus which he could use, driven by his own curiosity, experimental skills and uncanny insight, to identify and interpret important new research areas that have changed our view of the universe and its contents in a fundamental way.

K. I. Kellerman[2]

My determination to image the universe with high-angular-resolution instruments, although not shared by other experimenters and opposed on theoretical grounds by some theoreticians, drove me to push hard for the development and use of a telescope.

Riccardo Giacconi[3]

How Can Galaxies Be Imaged at Radio Wavelengths and in X-Rays?

Grote Reber (1911–2002) was "the first person who knowingly built a radio telescope."[4] This was in 1937. For almost ten years, he was also the only radio astronomer in the world. Reber was a creative engineer who explored multiple technologies to improve everyday life.

---

[1] M. Harwit, *In Search of the True Universe: The Tools, Shaping and Cost of Cosmological Thought*, Cambridge: Cambridge University Press, 2013, pp. 132–133.
[2] K. I. Kellerman, Grote Reber's Observations of Cosmic Static, in *The Astrophysical Journal: American Astronomical Society Centennial Issue*, 1999, Vol. 525, p. 372.
[3] R. Giacconi, *Secrets of the Hoary Deep, A Personal History of Modern Astronomy*, Baltimore: Johns Hopkins University Press, 1998, p. 114.
[4] P. Edwards, in *Biographical Encyclopedia of Astronomers*, T. Hockey, V. Trimble, and T. R. Williams (editors), New York: Springer, 2007, pp. 956–957.

Fig. 7.1 Grote Reber and his home-made 160-MHz radio receiver used to detect radio emission of cosmic origin. Credit: Image courtesy National Radio Observatory/AUI.

Reber had been trained as an electrical engineer with interests in ham radio (amateur radio communication) and astronomy. Intrigued by the pioneering radio work conducted at Bell Laboratories by Karl Jansky a few years earlier, he built a parabolic antenna 9 m in diameter that he erected in his home backyard in Wheaton, Illinois (Fig. 7.3). He used his facility to map the sky. He discovered important discrete radio sources, among others the strongest radio source in the sky, Cygnus A, making him the first to observe a radio galaxy. Always aiming for the simplest design, Reber stayed away from expensive instruments. He later lived and worked in Tasmania, Australia, where he studied the ionosphere and its effects on the propagation of low-frequency radio waves. We have reason to believe that Reber, with his sharp insight, knew that radio astronomy would develop into the thriving discipline it became. He lived to see the explosive growth of radio astronomy.

Riding the Waves or Multiwavelength Imaging

Evolution by natural selection has matched the sensitivity of the retina of our eye to the visible portion of the electromagnetic spectrum. This sounds like a tautology but this coincidence makes our eyes most sensitive to the wavelengths of the maximum sunlight in the spectrum. We call the eye's spectral range the optical domain, a wavelength range of about 0.4 to 0.75 micron, matching that of a thermal source with a temperature of about 5,700 K, close to the temperature of the solar surface.

In 1800, William Herschel carried out a slightly different version of Isaac Newton's experiment with sunlight. Letting sunlight through a prism and holding a thermometer just beyond the red part of the visible spectrum, William Herschel registered a rising temperature. He thus discovered that the Sun emitted "infrared radiation," beyond the red part of the visible spectrum. The radiation was invisible to the eye but detectable by other means. Astronomers have found over the course of the last 100 years that our own star, the Sun, emits much more than light and infrared radiation. These are small fractions of the whole electromagnetic spectrum, ranging from the shortest-wavelength radiation – gamma and X-rays during solar flares – to the longest, radio wavelengths, also produced during flare-associated disturbances. Physicists and astronomers have learned how to track these non-visible wavelengths and have created ways to detect them, leading to the development of innovative tools to "view" the sky at the different wavelengths. Using a variety of techniques, present-day astronomers register images over the whole electromagnetic spectrum. We can observe cosmic sources with almost all wavelengths: from the shortest wavelengths or highest frequencies – gamma rays and X-rays – to the longest or least energetic ones, the kilometric radio undulations. Because the Earth's atmosphere blocks the shortest wavelengths and the longer mid-infrared domain, it is necessary to get above the Earth's atmosphere to capture these emissions from cosmic sources; more on this at the end of the chapter.

Radio images of astronomical sources provide an extraordinary source of information. Several techniques were developed as the new field opened up. It has also been helpful that most radio waves can get through the Earth's atmosphere with minimal disturbance. They are not affected by atmospheric turbulence or by the presence of clouds. Because radio waves correspond to "long" photons, using them to make images has been very challenging. Meanwhile, the phenomenal onset of the Space Age in 1957 gave access to the short-wavelength domain by enabling the placement of telescopes and detectors above the Earth's atmosphere. Astronomers could take advantage of the stunning developments for detecting and recording shorter wavelengths, X-rays and gamma rays, which took place in the wake of World War II. These breakthroughs led to X-ray and gamma-ray astronomy. How did this happen in such a short period of time?

But let's begin with cosmic radio waves and look at the radio universe to find how astronomers learned to make images in the non-optical domains (see Plate 7.1).

From Cosmic Hiss to Radio Maps

The German physicist Heinrich Rudolf Hertz (1857–1894), who died tragically at the young age of 37, had set up experiments to produce electromagnetic waves and study them. In doing so, Hertz proved the existence of airborne electromagnetic waves and showed that they were propagated at the same speed as light. In his honor, these waves were called hertzian waves; now, they are simply known as radio waves. Hertz also discovered that shining ultraviolet light enhanced the production of sparks in the electrode gap of his receiver. This effect was the manifestation of the photoelectric effect that Albert Einstein would explain by quantum theory in 1905. The standard unit of frequency – cycles per second, hertz (Hz) – is named after him; one megahertz is written as 1 MHz. Hertz's findings turned out to be conclusive confirmations of James Clerk Maxwell's (1831–1879) electromagnetic theory of light. Soon after, other physicists found more links between light and electricity.

In the late nineteenth century, there were attempts to detect hertzian waves from the Sun; these were unsuccessful, as the Sun did not turn out to dominate the radio sky as it does in visible light. The first unambiguous detection of radio waves from the Sun was made by British physicist and astronomer James Stanley Hey (1909–2000), on 26, 27 and 28 February 1942, when investigating signals thought to be produced by the Germans to jam British radars.[5] In fact, radio waves had been registered from a far more distant source almost ten years before World War II, in a most unexpected way. The American radio engineer Karl Jansky (1905–1950) of Bell Laboratories was the first to detect cosmic radio waves in the early 1930s. It is a fine story.

The world was at the beginning of the radio communication age. Italian electrical engineer Guglielmo Marconi (1874–1937) had developed the first radio transmitters and receivers in the 1890s. Marconi shared the 1909 Nobel Prize in Physics with Karl Ferdinand Braun for the development of "wireless telegraphy." The propagation of radio waves was poorly understood. Jansky was investigating the atmospheric static that was affecting long-distance radio communications. Lightning from local storms or thunderstorms occurring in distant tropical regions produces a static radio noise that outbalances man-made signals. Jansky was searching for quieter periods (and directions in the sky) and wavelength windows that would improve and optimize transatlantic radio communications. He was also testing for directions of antenna where the reception of the man-made radio signal would be enhanced. However, there were other radio signals. In addition to the thunderstorm static that he recorded carefully, Jansky found a mysterious "hiss."

Observing at the long radio wavelength of 14.6-m (frequency of 20.5 MHz), Jansky established that the radio hiss had a periodicity of 23 hours 56 minutes. He noticed that this corresponded with the sidereal period of the Earth's rotation, the time it takes the Earth for its full rotation with respect to the stars. Certain enough of the celestial origin, Jansky referred to an *extraterrestrial* origin for this signal in his 1933 paper. By 1935, he had accumulated more data and was able to attribute the source of "interstellar interference" to the Milky Way.[6] The radio signal with the sidereal period was the strongest when his merry-go-round antenna pointed in the direction of the Sagittarius constellation where the center of the Milky Way, highest concentration of stars, is located (Fig. 7.2). For the first time in human history, electromagnetic signals from the cosmos other than light were being recorded.

Jansky's story is fascinating. Like several other discoveries in radio astronomy, his finding had been serendipitous. Two other remarkable serendipitous findings are the cosmic

---

[5] J. S. Hey, Solar Radiation in the 4–6 Metre Radio Wavelength Band, *Nature*, 1946, Vol. 157, pp. 47–48.

[6] K. G. Jansky, Electrical Disturbances Apparently of Extraterrestrial Origin, *Proceedings of the Institute of Radio Engineers*, 1933, Vol. 21, p. 1387; Karl G. Jansky, Note on the Source of Insterstellar Interference, *Proceedings of the Institute of Radio Engineers*, 1935, Vol. 23, pp. 1158–1163.

Fig. 7.2 **Transformational Image: Karl Jansky's Radio Map of the Sky.** Karl Jansky pointing to a portion of the Milky Way in the direction of the constellations of Cassiopeia and Cygnus where he detected enhanced cosmic noise shown as long wiggly lines. Credit: Image courtesy National Radio Observatory/AUI.

background radiation by Arno Penzias and Robert Wilson in 1964 and pulsars by the Northern Irish astronomer Jocelyn Bell and British astronomer Antony Hewish in 1967. As noted by British historian John North, "only gifted observers make chance discoveries."[7] The radio engineer Jansky was one of these gifted researchers and he had opened a new and extraordinary window into the universe. Within decades, many other sidereal objects,

[7] J. North, *Cosmos: An Illustrated History of Astronomy and Cosmology*, Chicago: University of Chicago Press, 2008, p. 661.

galaxies in particular, turned out to be rich sources of radio emission. Curiously, radio astronomy, as a new discipline of astronomy, would take a bit of time to emerge.

Radio Astronomy Slowly Gains Acceptance: Reber the Pioneer

Although publicized in the *New York Times* on 5 May, 1933, Jansky's discovery did not draw any attention from professional astronomers in the least.[8] His work remained in the shadows for a decade. The revelations of the great optical telescopes made in the 1920s and 1930s had been so transforming that "there was little reason to believe that significant contribution to our knowledge of the universe could come from other parts of the electromagnetic spectrum."[9] Nobody thought the radio domain would be significant for astronomy, and even less that astronomers would one day record spectacular images at radio wavelengths.

It was not until 1944 that the first radio astronomy publication appeared in a professional astronomy journal. This was the year Walter Baade had published his seminal paper, in the same volume of the journal, on the resolution of stars in the Andromeda Galaxy and its companions (Chapter 6). Grote Reber, who was not an astronomer but an electrical engineer and radio amateur, was the author of the groundbreaking paper. He reported on his mapping of cosmic static in the prestigious publication, *The Astrophysical Journal*.[10] This time astronomers did pay attention.

Reber's paper of 1944 had been the result of several years of patient work carried out in complete independence and isolation. Reber had built a 31-ft-diameter antenna with 2×4 lumber and metal sheets in his backyard, as well as the receiver equipment which he operated from his house (Figs. 7.1 and 7.3). With this rudimentary forerunner of future large radio telescopes, Reber produced maps of the radio sky at a wavelength near 1.87 m, much shorter than that used by Jansky. Reber observed at night to avoid the strong man-made radio interferences during the day, which came especially from automobile engine sparkplug firing.

Reber's work and findings were intriguing, challenging professional astronomers who were still on the defensive. In an unusual investigative move, the young astronomers Jesse Greenstein (1909–2002) and Gerard Kuiper (1905–1973) of the University of Chicago went to Reber's house in Wheaton. Following their visit and inspection of the facility, they reported back to Otto Struve, then editor of *The Astrophysical Journal*, at the University of Chicago. As faithful reporters in the field, Greenstein and Kuiper commented that the equipment "looked modern" and that Reber's work appeared "genuine."[11]

Reber's radio maps of the sky were in the form of isolines, which trace the constant-intensity levels in terms of 10^{-22} watts/square centimeter, per circular degree and per MHz band (Fig. 7.4). It is quite remarkable that Reber had carefully quantified the intensities (the

[8] R. Smothers, *Commemorating a Discovery in Radio Astronomy, The New York Times*, June 9, 1998.
[9] Sir B. Lovell, *The Story of Jodrell Bank*, New York: Harper & Row, 1968, p. 21.
[10] Grote Reber, Cosmic Static, *The Astrophysical Journal*, 1944, Vol. 100, pp. 279–287.
[11] Cited in K. I. Kellerman, Grote Reber's Observations of Cosmic Static, in *The Astrophysical Journal: American Astronomical Society Centennial Issue*, 1999, Vol. 525, p. 372.

Fig. 7.3 Grote Reber's radio telescope in his backyard at Wheaton, Illinois. Credit: Image courtesy National Radio Observatory/AUI.

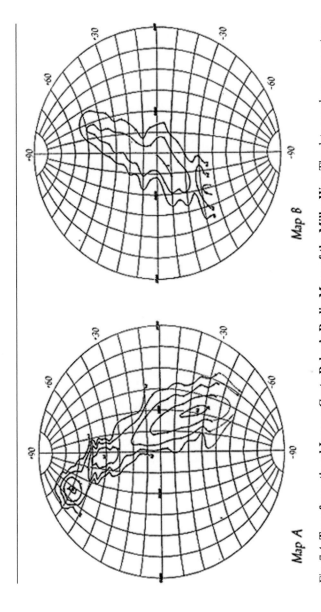

Map A   Map B

Fig. 7.4 **Transformational Image: Grote Reber's Radio Maps of the Milky Way.** The data are shown as contour maps. Credit: Image courtesy National Radio Observatory/AUI.

units must have baffled the optical astronomers of the time). The numbers on the central horizontal line indicate the right ascension (east–west direction) and the numbers around the circle perimeters, the declination (north–south direction). To emphasize the seriousness of his experiment, Reber included photographs of the radio dish, the power supply and the automatic recorder in the first two illustrations of his article. Then followed three pages of chart recordings of the 1.87-m radio signal corresponding to different declinations. The chart speed was 6 inches per hour as Reber's antenna moved only in declination.

Making Radio Waves

What caused the radio emission that was detected by Jansky and mapped by Reber? It is produced by free electrons moving through space at speeds close to the velocity of light as they spiral around and along the lines of the magnetic field threading its way through the tenuous interstellar gas clouds (Chapter 6). As the magnetic field and gas are concentrated in the plane of the Milky Way, it is then not very surprising that cosmic radio emission coincides with the luminous band of the Milky Way. It is also where the sources of energetic particles, novae, supernovae and evolved stars, are more abundant.

Electromagnetic radiation produced by relativistic electrons is called *synchrotron* radiation; relativistic refers to the speed of particles being close to the speed of light. This unusual type of radiation was identified in a General Electric laboratory accelerator in 1947.[12] A few years later, Swedish physicists Hannes Alfvén (1908–1995) and Nicolai Herlofson (1916–2004) showed that electrically charged particles trapped in the magnetic field surrounding a star could be accelerated in a similar way.[13] Later in the 1950s, well after Jansky's discovery and Reber's observations, Soviet astrophysicists Vitaly Ginzburgh (1916–2009) and Iosif Shklovsky (1916–1985) developed a detailed mechanism explaining synchrotron radiation and its properties.[14]

This non-thermal emission is distinct from the black-body emission that arises from any physical object emitting naturally by just being at a given temperature. Synchrotron radiation is a continuum of electromagnetic emission; the wavelength where it peaks in the spectrum depends on the energy of the particles and the strength of the field (Plate 7.2). The behavior of synchrotron radiation intensity as a function of wavelength is different from black-body radiation such as is emitted by the Sun. Hence, by making intensity measurements at various wavelengths, the synchrotron radiation can be identified from its characteristic spectral shape. Synchrotron radiation is also strongly polarized, that is, the observed intensity varies with the angle to the plane of the sky. Cosmic synchrotron radiation is observed most commonly in the radio region, but very energetic electrons spiraling in strong magnetic fields can also produce synchrotron emission in the visible domain and even X-rays. For example, powerful man-made light sources produce synchrotron radiation

[12] F. R. Elder, et al., Radiation from Electrons in a Synchrotron, *Physical Review*, 1947, Vol. 71, pp. 829–830.
[13] H. Alfvén and N. Herlofson, Cosmic Radiation and Radio Stars, *Physical Review*, 1950, Vol. 78, p. 616.
[14] I. S. Shklovsky, *Cosmic Radio Waves*, Cambridge, MA: Harvard University Press, 1960.

as very intense beams of visible light or X-rays that are used to sample all kinds of earthly materials.

Greber's pioneering work came at the time of huge developments in radar and electronics, accelerated by World War II. Benefiting from these developments and pushed by rising interest from professional electrical engineers and astronomers, the new discipline of radio astronomy blossomed. Surfing on the post-war technological wave, research teams successfully overcame several technical hurdles to use radio waves to explore our universe through a revealing new cosmic window and to make new kinds of images altogether.

Advantages and Limits of Radio Waves

Radio telescopes are analogous to optical telescopes in the way they operate. Radio waves coming from the sky are collected by an antenna, often but not necessarily a parabolic reflector, and sent to a focal plane where a receiver (instead of a photographic plate) registers them. Nonetheless, there are some important differences. Single-dish radio telescopes measure the electromagnetic signal originating from only one point at a time. Several pointing directions are required to make a meaningful radio map of a region of the sky. The image of an extended source is then assembled by joining points of equal brightness (isocontours) of radio signal intensity, as Reber did (Fig. 7.4).

Building radio images this way requires measurements of many contiguous spatial elements in the sky, each converted to a point measurement, a picture element, i.e. a pixel. The smaller the pixels, the finer the details of the image, and the greater the number of pixels, the larger the images are. Thus, making images or maps at radio wavelengths has for a long time required a huge amount of observing time and data reduction work. At the beginning of the book, I wrote: we read text; we read images. We also write text, and techniques of imaging with radio waves get close to what could be described as "writing" images. To overcome the obstacle of the time-consuming observing process for assembling images, radio astronomers had to come up with astute tactics. To understand how one gets around tedious point-by-point mapping, let us first review a few elements of wave physics.

A fundamental property of electromagnetic waves fixes the angular resolution that can be obtained using a given collector. For a lens, mirror or collector of a given aperture, the highest angular resolution, or finest detail, that can be resolved is set by the Raleigh criterion: $\theta = 1.22 \, \lambda/a$, where λ is the wavelength of the wave and a is the diameter of the collecting area, in the same physical unit, θ being the angular resolution expressed in radians. A *radian* is a unit of angular measure corresponding to the radius projected as an arc on a circle: one radian is 57.3 *degrees*. A 1-*degree* angle divides into 60 *arcminutes* (shortened to arcmin) where each arcmin splits into 60 *arcseconds* (shortened to arsec). For example, the human eye with a 3–9-mm variable pupil has an angular resolution of about 1 arcmin; for scaling purposes, the Sun and the Moon viewed in the sky are each about 30 arcmin in angular diameter. When using binoculars of, say, 8 × 35, which have lenses of 35 mm in diameter, an angular resolution of about 10 arcsec can be achieved; with this simple equipment, the main craters on the surface of the Moon can easily be distinguished.

The Hubble Space Telescope has a primary mirror that is 2.4 m in diameter; its resolution is about 0.1 arcsec, which is 600 times finer in angular size than that of the human eye.

From the Raleigh criterion it follows that large beam sizes result in less resolution. Radio telescopes of tens of meters in diameter appear huge, but they have beam sizes, i.e. resolutions, of a few arcminutes. This is very coarse because radio telescopes operate at much longer wavelengths than optical telescopes. Astronomers can detect and observe galaxies in the radio domain, but the large beam sizes (i.e. low resolution) of radio telescopes prevent structural studies of distant external galaxies.

In order to achieve higher angular resolutions, such as those obtained at optical wavelengths, collecting areas of tens of kilometers in size would need to be built, which is impractical and technically unachievable if one were to do this with a single monolithic parabolic dish.

Tricks for Making Finer Radio Images

To get around this fundamental limit, radio astronomers and engineers decided to mimic large apertures by a sort of technical cheat. With their usual knack, they invented an astute technique, *aperture synthesis*. The principle is that it is not necessary to work with filled or monolithic apertures, as with traditional parabolic mirrors; it is only necessary for the collecting surfaces to occupy patches of a large virtual area and to move them around. To understand the trick, let us make a detour into a few other fundamental aspects of imaging with electromagnetic waves.

Waves have interesting and useful properties. For example, waves of all sorts give rise to interference: amplifying when the peaks of the waves superpose (constructing interference) and canceling when peaks coincide with minima of the other wave (destructive interference). We observe this with water waves on a pond, "giant waves" in the ocean or sound waves in a room. Electromagnetic waves, light or radio, do the same. A multitude of set-ups, by way of an interferometer, create interference patterns that can be exploited.

A simple interferometer combines the signals of two antennae. The intensity of the waves (how strong it is) and their phases (where the wave is) need to be monitored. By analyzing the structure of the resulting signal after the interference, the angular size of a source, or the angular distance between two or multiple sources, can be estimated. Even the shape of a source can be inferred. It suffices to know the wavelength and the dimensions of the interferometer, i.e. the distance between its collecting elements and their orientation, with respect to the source under study. A more complex but also more powerful interferometer can be created by combining the beams of multiple antennae or radio telescopes. This has not been an easy task, but using this approach, brilliant individuals have managed to design and build a revolutionary type of imaging radio telescope. Obviously, the appropriate signal analysis technologies had to be in place.

In the 1960s, groups of British, Dutch, Australian and Canadian radio astronomers invented the technique by which a few radio telescopes, spread around but interconnected, could achieve the angular resolution equivalent to that of a large monolithic aperture; they called the technique *aperture synthesis*.[15] Instead of using a filled aperture or a very large single antenna, a number of distributed dishes or smaller antennae were positioned to form the multiple paired elements of a giant radio interferometer. Some of the telescopes could be moved on tracks, allowing pairs of different baseline lengths, overall mimicking a larger-sized telescope with holes in it, that is a partially filled surface. Signals received from each of the dishes were precisely synchronized to interfere. The arrival of atomic clocks enabled this high-precision work. There remained only the complex task of disentangling the interference patterns thus observed.

To do this, radio astronomers used very fast computers, called correlators, to compare the radio signals, record their phases and their intensities as received from a given point in the sky and for each separation and orientation of all sets of antenna pairs. To get more angular coverage, the rotation of the Earth was put to work, following an ingenious proposal and technique put forward by the British astronomer Martin Ryle (1918–84).[16] As the Earth rotates, the orientation of the different pairs changed with respect to the sky. Multiple pairs and the continuing change in their orientation generated many signals, and helped to reproduce a virtual surface close to a filled surface. Peter Scheuer has written a fine review of the early days of aperture synthesis.[17] However, organizing the multitude of interference patterns recorded for the different pairs, as orientations changed, requires more sophisticated mathematical tricks. A sort of "image" is obtained, but a further transformation needs to be applied, as one would need a specially wired brain to make up the real physical images from these raw data.

A Symphony of Waves

We can use a musical analogy to illustrate another important aspect of wave physics: harmonics. A musical symphony comprises the sounds of several instruments. By creatively combining the tunes of wind instruments, strings and percussion, composers, conductors and musicians produce wonderfully unified melodies. By letting the ear and the mind flow with the sounds, we forget the individual instruments. We are barely aware that they all contribute to the execution of the piece, coming in at different times and at varying sound levels. The temporal and intensity variations of the contributions by the groups of instruments are essential to the quality and impact of the final melody.

Just like musicians, mathematicians have found that natural shapes and phenomena (fixed or evolving with time) can be represented by the sum of relatively simple periodic trigonometric functions, sinusoids or cosines, or sine waves (Fig. 7.5). This is harmonic analysis. The simplest wave is the sinusoidal function that varies regularly over given cycles.

---

[15] W. T. Sullivan, III presents a fine review of early radio in Some Highlights of Interferometry in Early Radio Astronomy, in *Radio Interferometry: Theory, Techniques and Applications*, T. J. Cornwell and R. A. Perley (editors), Astronomical Society of the Pacific Conference Series, 1991, Vol. 19, pp. 132–149.

[16] Martin Ryle, The New Cambridge Radio Telescope, *Nature*, 1962, Vol. 194, pp. 517–518.

[17] P. A. G. Scheuer, The Development of Aperture Synthesis at Cambridge, in W. T. Sullivan, III (editor), *The Early Years of Radio Astronomy*, Cambridge: Cambridge University Press, 1984, pp. 249–265.

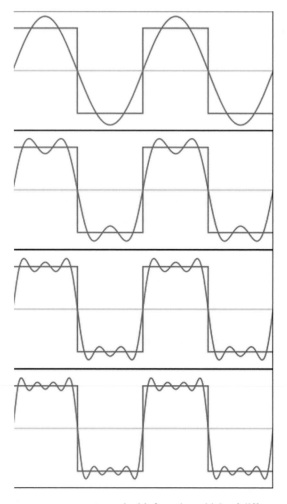

Fig. 7.5 A square wave reconstructed with four sinusoidals of different amplitudes.

To extract the contributing waves, the signal needs to be decomposed. The decomposition technique was explored by several mathematicians of the seventeenth and eighteenth centuries. However, it was French mathematician and physicist Jean-Baptiste Joseph Fourier (1768–1830) who introduced and applied the principle of oscillating functions to natural phenomena, first to study the propagation of heat in solid bodies.[18] It is an interesting aside that Fourier was also a mentor and strong supporter of the young Jean-François Champollion (1790–1832) who decrypted the texts on the Rosetta stone.

When you listen to a fugue or toccata from Johann Sebastian Bach, instruments enter a melody at different times; they are *phased* to create a new melody. The appropriate

---

[18] J.-B. J. Fourier, *Théorie analytique de la chaleur*, Paris: Chez Firmin Didot, père et fils, 1822.

gathering of various sinusoids can reconstruct a shape: the coefficients determine their relative strength, and their phase, i.e. the amount of shift between the various sinusoids.

As illustrated in Fig. 7.5, it is possible to reproduce any geometrical shape by combining in a weighted fashion an infinite number of sinusoids: it is a matter of adjusting the phases (that is shifting the waves with respect to each other) and experimenting with their peak intensities. This is the principle of a Fourier series. Fourier series and their more sophisticated representations, Fourier transforms, are the mathematical tools used to calculate or extract the *spectrum*, i.e. the frequencies and amplitudes of the set of sinusoids needed to reproduce a physical shape or a time variable signal. For example, the spectrum of sinusoids that reproduces a square box is the function sin $(x)/x$, also called a sinc (x) function, where x is a fractional or integer value of π.

Fourier series and Fourier transforms have become extremely powerful analytical tools in many fields. They are applied to a wide range of mathematical, physical and engineering problems, particularly in image processing and signal reconstruction. They play a part in just about every single high-technology device used today. So when you look around, remember anything can be reproduced just by a carefully designed sum of sinusoids.

Synthesis of Radio Images

Let us recap. Using several pairs of antennae or radio telescopes at varying separations, a cosmic source is sampled at different angular scales and resolutions. With N antennae, there are effectively $N(N-1)/2$ usable pairs. For example, the 27 antennae of the Y-shaped Very Large Array in New Mexico provide 351 simultaneous baselines (see Plate 7.3; see also the ALMA antennae, Plate 6.1). Letting the Earth's rotation change the orientation of the various pairs, the brightness pattern and phase structure can be mapped over different position angles in the sky. More specifically, it is as if you had assembled a good part of a very large reflector by strategically positioning your small antennae and letting the rotation of the Earth fill more and more of the area of the virtual large reflector.

Fast computations run on powerful correlators allow astronomers to construct maps of the amplitudes and phases of the radio signal received from all sets of pairs of antennae. Changing the distances between the pair allows the sampling of different spatial separations (i.e. angular resolution) and the Fourier plane (of frequencies and orientations) is filled with antennae positioned at as many different baselines and orientations as is operationally feasible. The whole set of sampled waves, the Fourier plane, can, in our analogy, be called the full symphony.

An interferometer of this sort, working for several hours, facilitates the measurement of the Fourier components of the radio brightness distribution, i.e. their spatial frequencies across a source or the angles at which dominant structures repeat themselves. It is not possible to get pairs of signals for an infinite number of antennae pairs, separations and orientations. Incomplete sampling is a fundamental but well-known problem of signal processing, analysis and reconstruction, as it is not possible to derive all the Fourier components. Mathematical tools have been developed to work around the limitation and to reconstruct images assuming that a minimum number of configurations are observed. When the observer has compiled enough Fourier components, he or she does a mathematical conversion called a Fourier transform to obtain a two-dimensional image. Amazing, indeed.

Imaging by aperture synthesis has improved enormously since the 1960s. Radio astronomers now produce astonishing radio images of astrophysical objects that compare in resolution with the best images of optical telescopes on the ground and in space. Furthermore, because radio antennae can be disposed around the surface of the Earth and even in orbit around it, the angular resolution obtained is the highest of all wavelength regimes, reaching the microarcsec scale, i.e. 100,000 times better than the best images of the Hubble Space Telescope. This is an amazing turn-around. However, optical astronomers have not lagged behind. They are also using interferometry and aperture synthesis. The signals of optical telescopes separated by a few hundred meters are combined in the same way as for radio interferometers. However, because of the blurring effect of our turbulent atmosphere, the technique suffers from limitations that affect radio waves to a lesser degree.

The arrival of aperture synthesis has been revolutionary. The 1974 Nobel Prize in Physics was awarded to British astronomers Martin Ryle (1918–1984) and Antony Hewish "for their pioneering research in radio astrophysics: Ryle for his observations and inventions, in particular of the aperture-synthesis technique, and Hewish for his decisive role in the discovery of pulsars." Apart from being visionary, Ryle was a very pragmatic scientist. Peter Scheuer recalled Ryle telling him about implementing the technique in the early days: "On engineering topics you shouldn't write mere theory, you should jolly well build the thing first."[19]

Imaging Cosmic Dragons

Early in the investigation of galaxies, it was found that the centers of some galaxies were sites of weird phenomena. Already in 1909, the German-born American astronomer Edward Arthur Fath (1880–1959) had noticed the presence of very strong and broad emission lines in the spectrum of the nucleus of the spiral galaxy NGC 1068. Fath was an astronomer at Mount Wilson Observatory. Allan Sandage writes that Fath left the observatory in 1913 because of a conflict with director Walter Adams. "Had he stayed, Fath would have become the Edwin Hubble of the observatory," adds Sandage.[20] Fath had found that spirals had spectra like star clusters, but the spiral NGC 1068 in the constellation of Cetus was one of the exceptions.[21] In 1926, Edwin Hubble added a few more galaxies exhibiting central activity. The American astronomer Carl Keenan Seyfert (1911–1960) carried out a whole study of this class of objects, hence the name "Seyfert galaxies."[22]

[19] P. A. G. Scheuer, The Development of Aperture Synthesis at Cambridge, in W. T. Sullivan III (editor), *The Early Years of Radio Astronomy*, Cambridge: Cambridge University Press, 1984, pp. 249–265.

[20] A. R. Sandage, *Centennial History of the Carnegie Institution, Volume 1: The Mount Wilson Observatory*, Cambridge: Cambridge University Press, 2004, p. 88.

[21] E. A. Fath, The Spectra of Some Spiral Nebulae and Globular Star Clusters, *Lick Observatory Bulletin*, 5, 1909, No. 149, pp. 71–77.

[22] C. K. Seyfert, Nuclear Emission in Spiral Nebulae, *The Astrophysical Journal*, 1943, Vol. 97, pp. 28–40.

Radio observations were key in revealing the nature of these strange sources. Cygnus A, first observed by Grote Reber in 1939, was the strongest discrete source of radio waves. The "radio star" was later identified as a giant elliptical galaxy at a distance of 700 million light-years. With aperture synthesis, it was shown to be surrounded by two giant radio lobes of synchrotron radiation, one on each side of the galaxy like a giant butterfly deploying wings. Cygnus A became the archetype of a fascinating family of extragalactic objects, radio galaxies (see Plate 7.4).

Indeed, the most impressive astrophysical phenomena seen at radio wavelengths are these giant radio galaxies: huge clouds of energetic particles, ejected by active galaxies emitting synchrotron radiation, spewed and ballooning out as colossal lobes of radio emission. The lobes appear to be roped right into the core or nucleus of the galaxy. The nature of the central sources or engines remains somewhat mysterious, but is probably explained as a massive black hole nurturing high-energy processes. Like a particle accelerator, the galaxy's central engine sends out energetic electrons and protons in jets extending sometimes to millions of light-years. The precessing jets fan out as spectacular lobes of synchrotron radiation. The majority of the sources are associated with elliptical galaxies, but some spirals also harbor an active galactic nucleus. Cygnus A, the most powerful radio source in our region of the universe, produces several million times more radio energy than a normal galaxy; its output corresponds to ten times the energy produced at all wavelengths by our Milky Way. Quasars are the most energetic and distant members of the active galactic nuclei family, reaching luminosities 100 times greater than that of the Milky Way.

Atoms in a Spin

Energetic electrons spiraling along cosmic magnetic field lines are not the sole sources of radio emission. Hydrogen, the most abundant element in the universe, is an important source of radio emission, through two different processes (spin transition and *bremsstrahlung*, discussed later), which greatly help in the understanding of the physics of cosmic matter. The first process is a very subtle atomic transition, which is responsible for the extraordinary radio spectral line at 21 cm (a frequency of 1,420 MHz). This spontaneous transition is of huge astronomical importance because it arises from neutral hydrogen, the most common form of the element in the universe; it is produced by atoms of hydrogen at temperatures in the range of a few to about 100 kelvins.

Let us draw in our mind a simple picture of a hydrogen atom: the positive proton is at the center of the atom, circled by the light negative electron. Imagine both the electron and the proton as little spinning tops. These miniature tops have a quantum property that allows their axes to be either aligned, i.e. their spinning axes pointing in the same quantum direction, or oppositely aligned – in the analogy of tops, with spin axes pointing in opposite directions. In the quantum world, these two configurations correspond to two different energy states. Amazingly, about once every few million years, the electron top flips spontaneously, and points in the opposite direction to what it was before. The minute flip produces a radio photon at a wavelength of 21 cm (1,420 MHz); it is the difference in energy between the two states of parallel versus opposite alignment. The transition is called "forbidden," meaning that the probability of the flip is extremely low, occurring only once every few million years. Most often the flip is triggered by collisions between hydrogen atoms which don't result in a radio photon, as the two colliders get away with the energy. However, because there is a lot of hydrogen in many low-density regions in galaxies, the 21-cm line is produced continuously throughout the universe, and can be observed at great distances.

The Race to the Spin

Dutch astronomer Konrad C. van de Hulst (1918–2000), then a Ph.D. student working with Jan Oort (1900–1992) at Leiden Observatory, predicted the spin transition of neutral hydrogen in a 1945 paper.[23] In 1949, Iosif Shklovsky (who had imagined a mechanism for synchrotron radiation) confirmed that the hydrogen radio line should exist and could be detected. The Dutch group searched for several years. It was Harold I. Ewen (1926–2013), then a brilliant young American astronomer and graduate student at Harvard University, working with Edward Purcell (1912–1997), who found it first.[24] The duo had developed a new method of radio observing called "switched frequency" mode. Using a quickly assembled horn antenna, made of plywood covered with copper foil mounted outside a laboratory window, they detected the hydrogen spectral line. They soon shared their technique with Dutch and Australian colleagues who almost immediately also found the line.[25] Neutral hydrogen, or HI emission, was ubiquitous.

The discovery was the start of another immensely fruitful episode in radio astronomy. The HI radio line became a superb tool to map the gas in the interstellar medium of our Milky Way and in nearby galaxies. The spiral structure and the rotation of the Milky Way was confirmed for the first time by HI observations conducted by Oort and his team (Fig. 7.6).[26,27] Following other successful observations by the pioneering teams, the radio telescopes being constructed and the existing ones were modified to better detect and measure the 21-cm line, and record radio images of galaxies and other objects.

A strong astrophysical driver for these facilities was to employ the 21-cm line to map the kinematics and dynamics of the cold hydrogen gas in galaxies and then use Newtonian physics to infer the total mass of galaxies. The HI line became most advantageous because it is intrinsically very narrow, due to the very long lifetime of the spin quantum energy states. Moreover, the interstellar medium is mostly transparent to this sharp radio line.

[23] H. C. van de Hulst, Radio Waves from Space: Origin of Radio Waves, *Nederlands tijdschrift voor natuurkunde*, 1945, Vol. 11, pp. 210–221.
[24] H. I. Ewen and E. M. Purcell, Radiation from Galactic Hydrogen at 1,421 MHz, *Nature*, 1951, Vol. 168, p. 356.
[25] C. Alex Muller and Jan H. Oort, The Interstellar Hydrogen Line at 1,420 MHz and an Estimate of Galactic Rotation, *Nature*, 1951, Vol. 168, pp. 356–358.
[26] J. H. Oort, Observational Evidence confirming Lindblad's Hypothesis of a Rotation of the Galactic System, *Bulletin of the Astronomical Institutes of the Netherlands*, 1927, Vol. 3, pp. 275–282.
[27] Jan H. Oort, Frank J. Kerr and Gart Westerhout, The Galactic System as a Spiral Nebula, *Monthly Notices of the Royal Astronomical Society*, 1958, Vol. 118, pp. 379–389.

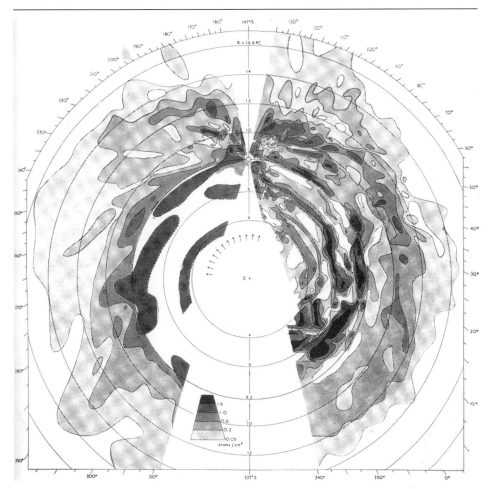

Fig. 7.6 **Transformational Image: The Milky Way Map of Neutral Hydrogen**. Distribution of neutral hydrogen in the plane of the Milky Way observed at 21 cm.[28] Maximum densities in the z-direction are projected onto the plane and contours of equal densities are drawn (darker is denser). The Sun is at the juncture of the axis, and galactic longitudes are indicated. The mapping of the structure was derived from the observed intensities and radial velocities of the 21-cm line along different lines of sight. The authors assumed circular symmetrical symmetry and used the model of the rotation curve derived by Maarten Schmidt.[29] Elongated features are mimicking spiral arms. This is a fine example of non-homomorphic representation (Chapter 4). From Oort, Kerr and Westerhout (1958), *Monthly Notices of the Royal Astronomical Society*.

[28] J. H. Oort, F. J. Kerr, and G. Westerhout, The Galactic System as a Spiral Nebula, *Monthly Notices of the Royal Astronomical Society*, 1958, Vol. 118, pp. 379–389.

[29] M. Schmidt, A Model of the Distribution of Mass in the Galactic System, *Bulletin of the Astronomical Institutes of the Netherlands*, 1956, Vol. 13, pp. 211–222.

With the advent of aperture synthesis in the 1960s, radio interferometers were built and used in Westerbock (Holland), Mullard (UK), Penticton (Canada) and Greenbank (USA). They revealed many outstanding features of the neutral hydrogen in galaxies. At the relatively small scale of a few light-years, the gas is distributed in filaments. On a larger scale, the spiral arms are seen in neutral hydrogen; they are regions corresponding to an increase of interstellar gas density. Within a few years, the research groups at Leiden and then Groningen, in Holland, had mapped the spiral structure of the Milky Way and were able to derive the rotation curve of the Milky Way inside the Sun's galactic orbit (Fig. 7.6).

Mapping neutral hydrogen became a "sport." It was quickly found that neutral hydrogen extended much further than the disks of galaxies seen in the visible light, sometimes twice, even three times, further (see Plate 7.5). Maps of the radial velocity of the neutral hydrogen were made of many galaxies (see Plate 4.2). But a mystery appeared. Instead of a falling off in a Keplerian way as a function of distance, which would indicate a tapering galaxy mass distribution with radial distance, the rotation velocity was found to remain constant to the edge of the neutral hydrogen disk (Chapter 8). In addition to rotation, peculiar structures like warps and non-symmetrical velocities were found, revealing morphological and kinematic anomalies. Most were assigned to interactions with companion galaxies, to mergers or to other non-axisymmetric behavior that is indicative of past cannibalism, where a galaxy has evolved by swallowing a companion or smaller satellite galaxies.

Images in the Submillimeter Range

The successful technique of aperture synthesis was also being applied to shorter radio wavelengths, for example in the millimeter wavelength range, where molecules, in particular CO, can be studied and molecular spectroscopy can be carried out. The Atacama Large Millimeter Array (ALMA) in northern Chile, with its 64 antennae, is the most recent and powerful of these new millimeter/submillimeter arrays, and is capable of imaging galaxies at very large distances (Plate 6.1).

But what about more direct radio imaging, as we do with our digital cameras? Surmounting severe technical challenges, radio astronomy did not lag behind other astronomy fields for long in developing panoramic electronic detectors. Over the last 20 years, we have witnessed the arrival of multi-pixel radio-wavelength detectors equivalent to the CCDs used for visible light. These arrays of detectors are capable of making direct radio images of the sky in the millimeter and submillimeter wavebands. The current technique uses arrays of miniature bolometers, which are lined up in rows and columns like a chessboard to register radio signals from a portion of the sky that is focused by a parabolic antenna of high surface precision.

A bolometer is a kind of thermometer. The radio photons are absorbed in each individual bolometer by a thin metal film cooled to a very low temperature, about 0.280 K, just above absolute zero. Heated by the sky's radio signal, the metal changes its temperature as it absorbs the radiation. A heat-sensitive semiconductor measures the change as a voltage blip, the voltage being proportional to the intensity of the incoming radiation. In current devices, the thermal, electrical and mechanical structure of the bolometer array is set on a single silicon wafer. Freestanding silicon nitride membranes, of less than one micron in thickness, are laid out with thin layers of titanium, which absorb the radiation. The temperature change in the absorbers is measured through semiconducting germanium chips, doped by a neutron transmutation, with electrical resistance that is acutely dependent on temperature. Making images with impinging radio waves is another amazing achievement of human ingenuity.

To complete this chapter on imaging outside the optical window, let us swing to the other end of the electromagnetic spectrum, the X-ray domain.

From Conic Sections to X-Ray Telescopes

Many regions in the interstellar medium are filled with hot gases of a few thousand degrees; these are the diffuse galactic nebulae; the Orion Nebula is the most spectacular such object nearby. The gas is ionized; it is called a plasma: atoms, mostly hydrogen and other heavier atoms such as oxygen, carbon and nitrogen, have lost some or most of their electrons due to ionization by ultraviolet radiation or collision with surrounding fast electrons. When the free electrons pass nearby protons or ions, the attractive electrical force decelerates them. While braking, the electrons lose some of their kinetic energy, which is converted into photons; this overall process is termed *bremsstrahlung*. This results in a broad continuum of wavelengths. When the gas is extremely hot, i.e. millions of kelvins, the braking emission peaks in the X-ray region. While synchrotron radiation from relativistic charged particles that are spiraling in magnetic fields is responsible for most of the radio spectrum at longer wavelengths, braking radiation, or *bremsstrahlung*, is the main source of emissions at the shorter radio wavelengths.

With the advent of the space age, astronomers saw a new opportunity opening: observing from outer space. They swiftly designed telescopes to be flown in Earth's orbit above the blocking atmosphere, gaining access to the shorter-wavelength domain, ultraviolet light, X-rays and gamma rays. For visible or infrared light, astronomers use lenses made of glass with a refractive index greater than one, or slightly curved glass mirrors. Light hits them at near normal angles and is redirected. To capture X-ray images of cosmic sources and image these, experimenters faced a challenge. X-ray photons have sufficiently high energies that they penetrate the material and are absorbed instead of being reflected by it, as light is by a metallic surface. To deflect X-rays, a completely new design of telescopic optics was required. The successful approach was to have X-rays hit a metallic surface at a shallow angle (from ten arcmin to a few degrees) and to deviate the rays only slightly, like a pebble being skimmed at the surface of a pond and bouncing up. This led to grazing-incidence metallic mirrors (Fig. 7.7).

An imaging X-ray telescope became a nested set of elongated metal cones following the principles outlined by German physicist Hans Woltjer (1911–1978) in 1952.[30] In

---

[30] H. Woltjer, Spiegelsystems streifenden Einfalls als abbildende Optiken für Röntgenstrahlen, *Annalen der Physik*, 1952, Vol. 445, pp. 94–114.

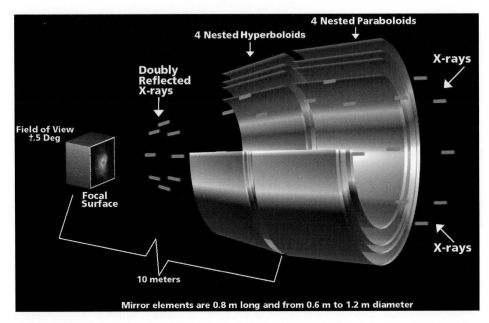

Fig. 7.7 Mirror elements of the Chandra X-ray Observatory showing the nested conic mirrors focusing X-rays on the detector. Credit: NASA, CXC, D. Berry.

his autobiography, American astronomer Riccardo Giacconi described how the twentieth-century collectors of X-rays had actually been designed in ancient times. While reading *Lo specchio ustorio, ovvero trattato delle sezioni coniche* (The burning mirror, or treatise on conic sections) published by the Italian mathematician Francesco Bonaventura (1598–1647) in 1632, Giacconi was struck by the drawing of Figure XIII in the ancient work. Stunned, he realized that he had replicated the principle of the collector that Bonaventura had conceived some 330 years earlier.[31] More startling, Bonaventura even referred in his own work to a book with the same title written by the ancient Greek mathematician Apollonius of Perga (c. 262–190 BC). Apollonius is noted for his work on conic sections, which influenced many scholars of the Renaissance and later periods. He had shown that the main geometrical curves (circles, ellipses, parabolae and hyperbolae, all names he created) were obtained by mentally pushing a plane through a cone and contemplating the figure of their intersection. So the twentieth-century technology for X-ray imaging is rooted in Antiquity.

X-ray astronomy has been an area of frontier work. The recent Chandra X-ray Observatory has been used to observe many clusters of galaxies and galaxies. The observatory was named for the Indian-born American astrophysicist Subrahmanyan Chandrasekhar (1910–1995) who obtained the Nobel Prize in Physics in 1983 for his mathematical theory of black holes. Then Giacconi was the recipient of the same award in 2002 for his work

---

[31] R. Giacconi, *Secrets of the Hoary Deep, A Personal History of Modern Astronomy*, Baltimore: Johns Hopkins University Press, 1998, pp. 137–138.

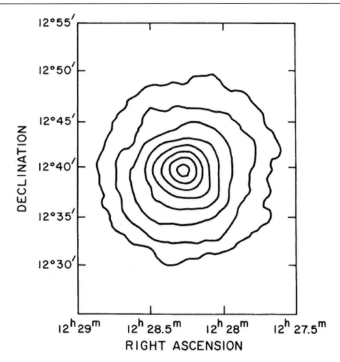

Fig. 7.8 **Transformational Image: X-ray Contour Map in the 0.7–3.0 keV energy band of the Giant Elliptical Galaxy Messier 87**. The intensity and radial distribution of the very hot gas indicates that the galaxy possesses a massive dark halo tens of trillions times the mass of the Sun. Credit: D. Fabricant, M. Lecar, and P. Gorenstein (1980), *The Astrophysical Journal*. © AAS. Reproduced with permission.

on cosmic X-ray sources, in particular clusters of galaxies, and for his designs for X-ray telescopes.

Measurements of the radio, optical and X-ray *bremsstrahlung*, of its shape and intensity, provide key diagnostics of the temperature and density of the interstellar gas. Observed from above the Earth's atmosphere, many stars, nebulae and galaxies have been found to be sources of X-rays (see Plate 7.6). Unexpectedly, the most spectacular sources of extended X-ray emission have turned out to be large elliptical galaxies and clusters of galaxies (Fig. 7.8). Clusters of galaxies are huge conglomerates, containing from hundreds to thousands of galaxies of all sorts, packed in a volume of a few million light-years in diameter. For example, the Coma cluster, well known to us from the time of William Herschel, hosts at least 1,000 bright galaxies (Fig. 1.4).

Except for a few cases, X-rays from the galaxy clusters do not come from the individual galaxies. The whole cluster volume is shining in X-rays, as if galaxies in these clusters are orbiting through a giant bath of extremely hot gas. The X-ray emission is produced by a colossal mass of ionized hydrogen and helium, with traces of heavy elements such as iron, at temperatures of tens of millions of kelvins (Plate 8.1). By mapping the neutral hydrogen of spiral galaxies in the inner parts of large clusters of galaxies, radio astronomers have shown them to be severely depleted of their neutral hydrogen (see Plate 7.7). As they revolve through the hot intracluster medium, cluster galaxies are shred, like trees losing leaves in a windstorm.

Strangely, the hot gas is gravitationally bound by the strong gravitational potential and is not boiling off the galaxy clusters. It has turned out that the intracluster gas is another indicator of some invisible mass or dark matter. Above all, clusters of galaxies imaged by X-rays have revealed more acutely than other astronomical objects the presence of dark matter on a large scale. The total masses of large galaxy clusters as derived from a range of indicators – stellar light, intracluster gas X-ray emission and galaxy kinematics – are huge. They are of the order of 10^{13} to 10^{14} solar masses. Let us now plunge into the topic of dark matter that has been mentioned so many times. Imaging with several techniques has been a key tool in revealing the presence of this mysterious entity.

8

Imaging the Invisible

Systems constructed by science resemble imposing edifices which, on closer inspections, reveal a crack or two.

Stanley L. Jaki[1]

In this darkness there is a small admixture, a few percent of the whole, which consists of the ordinary matter that makes up the stars and planets and us.

Steven Weinberg[2]

As I have shown previously, the probability of the overlapping of images of nebulae is considerable. The gravitational fields of a number of "foreground" nebulae may therefore be expected to deflect the light coming to us from certain background nebulae.

Fritz Zwicky[3]

How Can We Image the Invisible?

Horace Welcome Babcock (1912–2003) was an American astrophysicist who spent the greater part of his career at the Mount Wilson Observatory (Fig. 8.1). He was the son of astronomer Harold Delos Babcock (1882–1965), a spectroscopist and solar physicist, who joined George Hale at the new Mount Wilson Observatory in 1909, and was the first to measure the magnetic field of the Sun. Horace was fascinated by instrument building and invented several new techniques for astronomical observing. In 1953, he pioneered adaptive optics, a technology now used in almost all large ground-based telescopes, which corrects atmospheric turbulence and provides images with close to diffraction-limit angular resolution.[4] Babcock also led and carried out the construction of a large observatory at Las Campanas, Chile, to survey the southern-hemisphere sky and to record images of

---

[1] S. L. Jaki, *The Milky Way, An Elusive Road for Science*, New York: Science History Publications, 1972, p. 291.

[2] S. Weinberg, Physics: What We Do and Don't Know, *The New York Review of Books*, 2013, Vol. LX, No. 17, p. 86.

[3] F. Zwicky, On the Masses of Nebulae and of Clusters of Nebulae, *The Astrophysical Journal*, Chicago, 1937, Vol. 86, pp. 237–238.

[4] H. Babcock, The possibility of compensating astronomical seeing, *Publications of the Astronomical Society of the Pacific*, 1953, Vol. 65, pp. 229–236.

Fig. 8.1 Robert Leighton (left) and Horace W. Babcock (right), two of the most creative scientists of the twentieth century. Credit: Courtesy of the Archives, California Institute of Technology.

galaxies of unprecedented quality. The Chilean site hosts several large optical telescopes, including the 2.5-m Irénée du Pont telescope, the source of thousands of images of galaxies, many used in the splendid atlases of galaxies that are presented in Chapter 10. Allan Sandage called Babcock the quiet American. "Horace Babcock's reluctance to advance his own agenda kept him from achieving the renown he deserves as one of the great minds of his generation. On the other hand, his inability to indulge in self-aggrandizement made him an extremely effective director of the joint Mount Wilson and Palomar Observatories."[5] Babcock provided one of the earliest pieces of evidence for the presence of a large amount of dark matter in spiral galaxies.

Invisible Mass

Astronomers had to swallow an uncomfortable surprise around the middle of the twentieth century. They found that most matter in the universe might be of a kind that neither absorbs nor emits radiation. They called the stuff "dark matter," but it would be more appropriate, as suggested by the American physicist Steve Weinberg, to describe it as transparent matter or invisible mass. Despite their extraordinary number and variety of shapes, and the amount of matter they contain, galaxies constitute only the tip of the iceberg of cosmic mass; they are islets in a sea of invisible matter. The invisibility is not because the material is very dark

---

[5] A. R. Sandage, *Centennial History of the Carnegie Institution of Washington, Volume 1: The Mount Wilson Observatory*, Cambridge: Cambridge University Press, 2004, p. 420.

or extremely cold; it is due to its unusual – and yet unknown – nature. Despite multiple efforts, we have not yet succeeded in breaking the Gordian knot of what dark matter really is. Antimatter is not dark matter: it is like ordinary matter; it emits and absorbs light. We can see and are able to create antimatter in the laboratory.

Nevertheless, we do know a few things about dark matter. First, it exercises a gravitational pull, just like ordinary matter; it creates gravitational potentials giving weight to objects and accelerating masses that fall into them. We know also that this invisible mass is, on average, about six times more abundant than ordinary matter, i.e. baryonic matter. The latter is made of electrons, protons, neutrons and of other exotic well-known species of particles such as neutrinos, mesons or bosons. Dark matter is none of this stuff. In this chapter, I will show how observations and images of ordinary matter have revealed the existence of this perplexing and ubiquitous component of our universe.

A Brief History of Dark Matter

Early in the twentieth century, astronomers used the absolute luminosity of stars and stellar systems, or the amount of light they emit, to derive the amount of matter contained in them. They devised clever ways to translate the amount of light coming from stars into the mass responsible for its emission. They also established how this ratio of "mass to light" varied for different types of stars. For example, low-mass stars produce relatively little light compared to massive stars, which emit copious amounts of light. British astrophysicist Arthur Stanley Eddington (1882–1944) found a simple relation between luminosity and mass: the luminosity of a star is proportional to approximately the power 3.5 of its mass, i.e. $L \sim M^{3.5}$.[6] A star five times as massive as the Sun is 280 times more luminous than our luminary. The Eddington relation applies to stars on the main sequence, those in the stable hydrogen-burning phase. To get the overall mass of a galaxy, one also had to estimate independently the mass of matter that emits very little light: very cool stars, dwarf stars, black holes, small solid bodies and cold interstellar gases. Applying Newtonian mechanics, the global kinematics of bodies in a galaxy provided an independent clue about its total mass and spatial distribution. Most puzzling, the results from the two methods disagreed very significantly. Something was unbridled.

Evidence for some dark matter was first inferred in the early 1930s. It came first from the measurements of velocities of nearby stars and gas clouds in our neighborhood of the Milky Way. The Dutch astronomer Jan Oort (1900–92) had been analyzing the stellar kinematics in the solar neighborhood to derive a mean cosmic mass density. In 1932, he found that this density was twice that deduced from the stars and the gas.[7] Compared with the results of the mass-to-light ratio analysis just discussed, the kinematics led to significantly more mass than was implied by the stellar light striking our telescopes. The velocities of the stars in

[6] A. S. Eddington, On the Relation between the Masses and Luminosities of the Stars, *Monthly Notices of the Royal Astronomical Society*, 1924, Vol. 84, pp. 308–332.

[7] J. H. Oort, The Force Exerted by the Stellar System in the Direction Perpendicular to the Galactic Plane and Some Related Problems, *Bulletin of the Astronomical Institute of the Netherlands*, 1932, Vol. VI, No. 238, p. 249–287

our neighborhood of the Milky Way were much higher than could be accounted for solely from the masses of visible stars, gas and dust clouds, even when all these masses were carefully summed up. To explain the discrepancy, Oort suggested the existence of an important component of invisible mass, or "missing mass." Studying the edge-on galaxy NGC 3115 in 1940, he also found that "the distribution of mass in this system appears to bear almost no relation to that of light."[8] Oort's interpretation of these data was highly debatable.

The puzzle grew when the curious kinematical behavior was found on a much larger scale, that of clusters of galaxies. Our old friend Fritz Zwicky (1898–1974) of Chapter 3 was an early and astute observer of the extragalactic world (Fig. 3.11). He had taken a keen interest in the regions of space where galaxies appear to congregate in volumes of only a few million light-years in diameter. As we have seen, William Herschel and the Birr Castle observers had noticed such groupings of galaxies in the nineteenth century but never speculated on the nature of these assemblies of "nebulae" except in describing them as a "stratum." We know today that they are galaxy clusters; galaxies group together as hundreds, sometimes thousands of members, to make up the so-called "rich" clusters. The American astronomers Harlow Shapley and Adelaide Ames were the first to come up with the concept of clusters of galaxies, calling them "nebular clouds."[9]

By 1933, Zwicky had prefigured the presence of an excess of mass in many clusters.[10] Using the newly installed wide-field 18-inch Schmidt telescope of Mount Palomar Observatory, Zwicky surveyed and catalogued thousands of clusters of galaxies. Employing the larger Mount Wilson 100-inch telescope for follow-up observations, he obtained the spectra of several of the individual galaxies of clusters, deriving their velocities with respect to the center of the cluster. Applying Newton's law of gravitation, Zwicky derived the masses of the entire cluster by measuring the velocities of the individual member galaxies, just as we can calculate the mass of the Sun by measuring the orbital speed of the planets and their distance to the Sun. As Jan Oort had found on a smaller astronomical scale, Zwicky discovered that the mass from the luminous material of the cluster was clearly insufficient to account for the high velocities he measured for the galaxies; these should have flown away. Surmising that they were being held together, he claimed that there was an important amount of "missing mass."

Zwicky had also patiently measured the rotational curves of a few clusters of galaxies to infer their individual mass from Newtonian mechanics. He summed together all their masses. The "missing mass" did not go away; the masses of the constituent galaxies did not add up at all to the total mass. The prescient Zwicky wrote, "if this is confirmed, we would arrive at the astonishing conclusion that dark matter is present with a much greater density than luminous matter."[11] Soon after, in 1936, the American astronomer and inventor

---

[8] J. H. Oort, Some Problems Concerning the Structure and Dynamics of the Galactic System and the Elliptical Nebulae NGC 3113 and 4494, *The Astrophysical Journal*, 1940, Vol. 91, pp. 273–306.

[9] H. Shapley and A. Ames, A Study of a Cluster of Bright Spiral Nebulae, *Harvard College Observatory Circular*, 1926, Vol. 294, pp. 1–8.

[10] F. Zwicky, Die Rotverschiebung von extragalaktischen Nebeln, *Helvetica Physica Acta*, 1933, Vol. 6, pp. 110–127.

[11] Translated from F. Zwicky and cited by S. van den Bergh, The Early History of Dark Matter, *Publications of the Astronomical Society of the Pacific*, 1999, Vol. 111, p. 657.

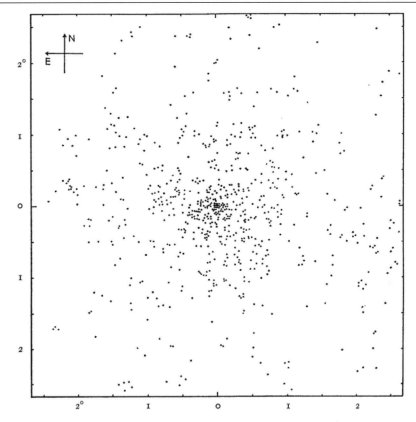

Fig. 8.2 **Transformational Image: Zwicky's Map of the Distribution of Galaxies in the Coma Cluster of Galaxies**. From Zwicky (1937), *The Astrophysical Journal*. © AAS. Reproduced with permission.

Sinclair Smith (1899–1938) also found the presence of large amounts of invisible mass in another nearby large galaxy cluster, the Virgo cluster of galaxies at about 54 million light-years.

In a fascinating paper in 1937, Zwicky summarized the extensive set of observations he had conducted of the Coma cluster of galaxies hosting hundreds of galaxies and located at about 333 million light-years (Fig. 1.4). While mapping the distribution of galaxies in the cluster, Zwicky had been struck by its apparent spherical symmetry (Fig. 8.2). He demonstrated that the galaxies in the Coma cluster were evenly distributed and that individual galaxies had much higher velocities than expected, based solely on the gravitational mass inferred from the stellar light of the galaxies.[12] Repeating Oort's remark about the solar

[12] F. Zwicky, On the Masses of Nebulae and of Clusters of Nebulae, *The Astrophysical Journal*, Chicago, 1937, Vol. 86, pp. 217–246.

neighborhood in our Milky Way, Zwicky concluded that light was not a good tracer of mass. Highly intrigued, he explored alternative approaches to calculate the total mass of clusters of galaxies in order to put his unexpected finding on solid ground and to find possible inconsistencies. In order to do this, Zwicky used an extraordinarily powerful tool, the *virial theorem*.

The Virial Theorem

The virial theorem is based on Newton's gravitational theory, and is a simple mathematical relation between the mass of a system, its size and the velocities of its components. It can be employed to infer the mass of a system of particles in hydrostatic equilibrium. Zwicky made the assumption that galaxies of a large cluster were particles of such a system. Taking the mean velocity of all the individual galaxies, σ, and the size of the cluster, R, the total mass, M, derived from the virial is written in its simple form as $M_{virial} \sim 5\,R\sigma^2/G$, where G is the gravitational constant. The basic assumption is simple: the system, in this case the galaxy cluster, has had the time to settle dynamically. The members are neither in free fall, nor escaping from the gravitational potential of the cluster. The beautiful morphological symmetry of the many galaxy clusters Zwicky had photographed indicated that the hydrostatic state assumption was reasonable, at least in the case of the Coma cluster.

Applying the virial equation, Zwicky arrived again at an astounding value for the total mass of the cluster. Combining the results of his several approaches, Zwicky announced the presence of a "missing mass" that was 100 times greater than that of the luminous matter. Zwicky was adamant: a kind of matter other than the luminous one was necessary to keep the cluster gravitationally bound, and to explain the smooth distribution of galaxies around its center.

Unsatisfied, Zwicky explored the possibility that clusters of galaxies were dynamically unstable or had not had time to settle. The galaxy velocities would then be unreliable tracers of the total mass and the virial theorem could not then be applied. He rejected this hypothesis, stating that clusters of galaxies would be so short-lived that there would be very few clusters in existence. This was not the case as clusters of galaxies peppered the deep sky. Zwicky was finding clusters all over, above and below the plane of the Milky Way, directions where he could probe deep into the universe. Most galaxy clusters must have had the time to relax gravitationally, making the use of the virial theorem justifiable and its results reliable, he re-affirmed.

Zwicky could not escape the stunning conclusion that clusters of galaxies were big pockets of dark matter, loci with huge concentrations of this matter that create deep depressions in the membrane of spacetime. Ordinary matter just falls and assembles into these deep "valleys." We know today that dark matter stretches across space as large-scale cosmic filaments; the filaments form a web of structures that crisscross each other; clusters of galaxies are found at the junctions of these large-scale filaments. The visible structures, i.e. the galaxies, groups of galaxies and clusters of galaxies, trace the dark matter as streetlights outline the patterns of our cities seen at night.

The Secret of Symmetry of Galaxy Clusters and X-Rays

When later observed from space, the clusters of galaxies revealed another surprise: they contain colossal quantities of X-ray-emitting gas. X-ray-emitting plasma from the Coma cluster, so well studied by Zwicky, was discovered during a balloon flight in 1966; this finding of hot intracluster gas was one of the major breakthroughs of X-ray astronomy (Chapter 7). Because of their deep gravitational potential, clusters of galaxies are able to retain and heat their intracluster gas to very high temperatures (Plate 8.1). In Coma, X-rays indicated a binding mass of about 7×10^{14} solar masses, with its intracluster gas heated to 100 million kelvin.[13] These high temperatures meant that the speeds of the ions were the same as those of the galaxies moving within the cluster's gravitational potential. Therefore, galaxies behave just like atoms in the deep potential created by the mass of the dark matter. Neither could escape. And the more massive the cluster was, the hotter the gas was found to be.

Even today, the origin of the gas remains a mystery, especially because it exists in such huge amounts. The suspicion is that it was produced by the explosions of past generations of massive stars located within the member galaxies of the cluster: the expelled gas from these galaxies dispersed into the surrounding medium, distributing itself evenly around a central condensation. The X-ray-emitting gas shows more spherical symmetry than the distribution of the individual galaxies. It maps the dark matter more closely than the galaxies. X-ray imaging from space has become the way to image indirectly the invisible mass in the universe. It is safe to say that imaging the X-ray emissions of galaxy clusters is imaging the dark matter distribution. It is one of the most direct ways to "see" dark matter, at least visualize its presence, and the extent to which it carves the deep gravitational potentials of clusters into the fabric of spacetime.

Rotational Curves of Disk Galaxies

Not too dissimilarly from Jansky's discovery of cosmic radio emission, "Zwicky's findings and arguments were not much noticed. It took about forty more years until observational evidence clearly showed that large amounts of invisible matter, gravitating matter was lodging in the disks of rotating galaxies. Without it, the galaxies would fly apart because of a too weak gravitational attraction."[14]

However, another prescient young astronomer had opened Pandora's box of dark matter. In his 1939 Ph.D. thesis work on the rotational velocity of the Andromeda "nebula," Horace Babcock (1912–2003) had found abnormally high rotational speeds in the outer parts of the Andromeda Galaxy (Fig. 8.3).[15] While Babcock refrained from making a link with the

---

[13] Plasma physicists express temperature in electronvolts (eV) or kilo-electronvolt (keV), where 1 eV = 11,605 K.

[14] H. S. Kragh, *Conceptions of Cosmos – From Myths to the Accelerating Universe: A History of Cosmology*, Oxford: Oxford University Press, 2007, p. 214.

[15] H. W. Babcock, The Rotation of the Andromeda Nebula, *Lick Observatory Bulletin*, 1939, Vol. XIX, No. 498, pp. 41–51. Plate III shows the image of M31, the points where spectroscopic measurements were obtained and a plot of the galactocentric velocities. Babcock concluded his extraordinary article comparing the kinematics of Messier 31 with that of the Milky Way.

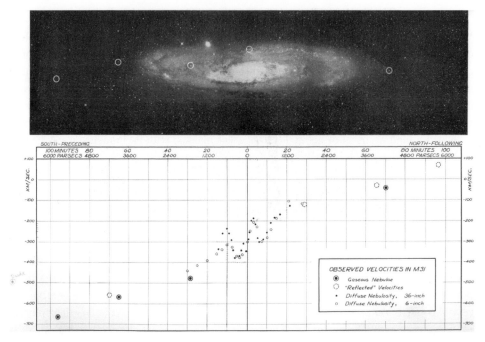

Fig. 8.3 Horace Babcock's rotation curve of the Andromeda Galaxy from his Ph.D. thesis, based on spectroscopic measurements with the Crossley 36-inch telescope. The most distant galactocentric points were derived from gaseous nebulae. From Babcock (1939), *Lick Observatory Bulletin*.

invisible mass inferred by Zwicky in the Coma cluster, he warned in the same way about the uncertainties of deriving masses of "nebulae" from spectroscopic rotational curves. Probably less bold than Zwicky, the young Babcock invoked absorption of light increasing outward from the nucleus, or some "internal gravitational viscosity" to be possibly greater than supposed and to produce the unexpected high orbital velocities of stars in Messier 31. A few decades later, radio observations of neutral hydrogen at 21 cm in M31 indicated that rotational speeds remained constant instead of falling in a Keplerian fashion as one gets further and further away from the center of the galaxy. This finding, 36 years after Babcock's noteworthy discovery, came from the American astronomers Morton S. Roberts and Robert N. Whitehurst who published their radio work in 1975.[16]

"These observations indicated that the mass in the outer regions of the Andromeda galaxy increased with galactocentric distance, even though the optical luminosity of M31 did not."[17] Sidney van den Bergh noted that neither Babcock nor Roberts and Whitehurst

---

"A new discrepancy is now directly apparent when the rotations of the two systems are compared, for the nearly constant angular velocity of the outer parts of M31 is the opposite of the 'planetary type' of rotation believed to obtain in the outer parts of the Galaxy."

[16] M. S. Roberts and R. N. Whitehurst, The Rotation Curve and Geometry of M31 at Large Galactocentric Distances, *The Astrophysical Journal*, 1975, Vol. 201, pp. 327–346.

[17] S. van den Bergh, The Early History of Dark Matter, *Publications of the Astronomical Society of the Pacific*, 1999, Vol. 111, p. 660.

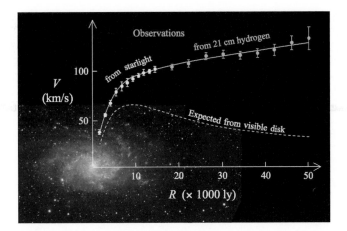

Fig. 8.4 The rotational curve of the spiral Messier 33. Credit: Stefania Deluca.

cited the 1933 or 1937 papers by Zwicky and his inference of dark matter.[18] Instead, as Babcock had hypothesized, Roberts and Whitehurst attributed the hidden mass to an abundance of dwarf M stars, the most common type of stars in the solar vicinity. These pioneers remained oblivious to the elephant in the room despite their highly reliable measurements.

In 1970, the Australian astronomer Kenneth Freeman suggested "undetected" and "additional" matter beyond the optical extent of several nearby galaxies.[19] A young American female astronomer and her colleagues picked up the thread three decades after Babcock's early hint. Studying the behavior of the rotational curves (the velocity of rotation as a function of distance from the galaxy center) in the 1970s, American astronomer Vera Rubin (1928–2016) and her colleagues showed, for several spiral galaxies, that their masses were many times that of their visible stars and gas content (Fig. 8.4). The team established compelling evidence for the presence of a large amount of dark matter in individual galaxies. They also found that the invisible mass was much less concentrated than visible light. "By the late 1970s dark matter had been discovered, not just hypothesized, and in amounts much larger than visible matter."[20]

James Binney and Michael Merrifield warned, cautiously, "the emergence of a tenuous distribution of utterly dark material cannot be ascertained by studying circular-speed curves and photometric profiles alone."[21] Dark matter was indeed difficult to swallow, as most astronomers and, at first, physicists had ignored it. In an interview, Vera Rubin stated: "I think many people initially wished that you didn't need dark matter. It was not a concept that people embraced enthusiastically. But I think that the observations were undeniable enough

[18] S. van den Bergh, The Early History of Dark Matter, *Publications of the Astronomical Society of the Pacific*, 1999, Vol. 111, p. 660.
[19] K. C. Freeman, On the Disks of Spiral and S0 Galaxies, *The Astrophysical Journal*, Vol. 160, pp. 811–830.
[20] H. S. Kragh, *Conceptions of Cosmos – From Myths to the Accelerating Universe: A History of Cosmology*, Oxford: Oxford University Press, 2007, p. 214.
[21] J. Binney and M. Merrifield, *Galactic Astronomy*, Princeton: Princeton University Press, 1998, p. 510.

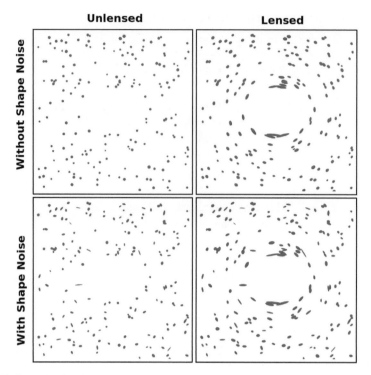

Fig. 8.5 Weak gravitational lensing inducing the distortion of galaxy shapes, assumed here to be ellipses. The effect shown is exaggerated relative to real systems. Credit: Wikipedia Commons/ TallJimbo.

so that most people just unenthusiastically adopted it."[22] Moreover, theoretical astrophysicists needed it to stabilize galaxy disks. Without dark matter, spiral galaxies would just fly apart like broken flywheels.

Imaging with Gravitational Lenses

As predicted by general relativity, concentrations of mass between us, as cosmic observers, and distant sources curve the fabric of spacetime. This curvature is capable of bending light paths. If there is a distant light source beyond a large mass like a cluster of galaxies, the interfering mass acts like a lens: it amplifies the apparent brightness of the distant source and distorts its shape. This weird phenomenon is called lensing, or mirage (Plate 8.2).

When the lensing effect is important, we have *strong* lensing. In its more subtle manifestation, it introduces small distortions in the shape of the object and is called *weak* lensing (Fig. 8.5). Although posited by our visionary Fritz Zwicky in 1937, the large-scale lensing

[22] Interview with Dr. Vera C. Rubin by Alan Lightman, Washington, April 3, 1989, Niels Bohr Library Archives, Center for History of Physics.

by spacetime curvature was first observed only in 1979.[23] With today's telescopes and their fine imaging capabilities, gravitational lensing is commonly observed. The Hubble Space Telescope and large ground-based telescopes have produced spectacular images of many such cases. As predicted by Zwicky, the most dramatic lensing effects are those created by large clusters of galaxies.

Gravitational lensing is an independent and powerful way of mapping dark matter in galaxies and clusters of galaxies and for viewing the most distant galaxies. Again, it was Zwicky who suggested in his seminal 1937 paper that gravitational lensing might be used to "weigh" galaxies. Distant galaxies viewed through the field of a midway galaxy cluster will have their shape distorted into elongated arcs. The amount of distortion is a measure of the mass of the cluster.[24]

Strong lensing is active through the inner part of the cluster and weak lensing in its outer parts. In gravitational optics, mathematical models are applied to reconstruct the shape of the distant galaxies and to estimate the amount of brightness magnification. As an inverse mathematical solution, the reconstruction spills out the total mass of the cluster. Relativistic physics applied to images of lensing clusters offers an elaborate and powerful way to image dark matter. By carefully mapping the distorted arcs, astronomers have been able to estimate not only the mass of the lens, i.e. the mass of the intervening cluster of galaxies, but also its geometry and spatial distribution. Lensing measurements and gravitational-potential reconstructions have confirmed that the major part of the mass of galaxy clusters is dark matter. The amounts derived are consistent with those inferred from the virial and X-ray measurements.

Moreover, large concentrations of dark matter in clusters can be employed as gravitational telescopes. Their light-amplifying power is exploited to probe the distant universe. Consequently, the lensing of clusters of galaxies allows us to look beyond and see the most distant galaxies in the universe, reaching even the youngest objects, born soon after the Big Bang. It is amazing that we use pockets of invisible matter as telescopes to look and image further in space and time. So, we know where dark matter is, how much there is, but we do not know what it is.[25]

This chapter on dark matter concludes the overview of modern ideas about galaxies and of how they developed using a whole range of imaging techniques and tools. Let us shift to a more targeted topic, the use of images in the pursuit of the world of galaxies: atlases of galaxies, their history as books of discovery and as tools for unveiling galaxies and learning about them.

---

[23] D. Walsh, R. F. Carswell and R. J. Weynmann, 0957+561 A, B: Twin Quasi-Stellar Objects or Gravitational Lens?, *Nature*, 1979, Vol. 279, pp. 381–384.

[24] The angle of deflection θ is given by $\theta = 4\,GM/(rc^2)$ toward the mass M at a distance r from the affected radiation; G is the gravitational constant and c the velocity of light.

[25] There are several excellent books on dark matter. For example, K. Freese, *The Cosmic Cocktail, Three Parts of Dark Matter*, Princeton: Princeton University Press, 2014; L. Randall, *Dark Matter and the Dinosaurs: The Astounding Interconnectedness of the Universe*, New York: HarperCollins Publishers, 2015.

Part III

Organizing the World of Galaxies

9

The Galaxy Classification Play-Off

The establishment of an ordering of galaxies into taxonomic categories is
clearly only one step towards understanding the physical nature of galaxy
structure.

Ronald J. Buta[1]

It is of supreme importance to the creation of our operational morphologies that we use the "things themselves" as our picture-language for stellar spectra, and direct photographs for galaxies and clusters of galaxies.

W. W. Morgan[2]

If the original Hubble scheme were to be called a Mark I model, the
revised but unpublished Hubble–Sandage system might be regarded as
Mark II, and the writer's present system as Mark III.

Gérard de Vaucouleurs[3]

Is the World of Galaxies too Chaotic for a Meaningful Classification?

Adelaide Ames (1900–1932), the Pickering Fellow in 1923, was the first woman graduate
student in astronomy at Harvard University (Fig. 9.1). She worked on "bright spiral nebulae" with the ebullient Harvard College Observatory director Harlow Shapley (Fig. 5.2).
Ames is seen on some of the photographs picturing life at the observatory in the 1920s and
1930s. She died tragically in a canoe accident at Squam Lake, New Hampshire, on June 26,
1932, the year of the publication of the Shapley–Ames catalogue of bright galaxies of which
she was co-author. She was probably one of the first astronomers to have had a full, though
too short, career devoted entirely to the study of galaxies. She worked on the properties of
galaxies and of clusters of galaxies and found deviations from isotropy in the general distribution of galaxies. Ames had been appointed to the Commission of Nebulae and Clusters at
the International Astronomical Union Assembly of Leiden, the Netherlands, in 1928. One
wonders if Ames foresaw how explosively the field of galaxy research would develop just

---

[1] Ronald J. Buta, http://ned.ipac.caltech.edu/level5/March01/Buta/Buta4.html
[2] W. W. Morgan, A Morphological Life, *Annual Review of Astronomy and Astrophysics*, 1988, Vol. 26, pp. 1–10.
[3] Gérard de Vaucouleurs, Classification, Dimensions, and Distances of Bright Southern Galaxies, *Sky and Telescope*, 1957, Vol. 16, p. 582.

Fig. 9.1 Adelaide Ames. Credit: Schlesinger Library, Radcliffe Institute, Harvard University.

in the few years following her untimely death. Classifying galaxies was a key feature of the new catalogue she and her supervisor had put together.[4] This catalogue would become the roadmap and guide of research on galaxies for the remainder of the twentieth century.

Atlases as Reference Visual Displays

On removing oneself a little from the multitude of objects and physical components, it is possible to find order in the jungle of galaxies. In contrast to gaseous nebulae, chaotic processes are present but rarely dominate the morphology of a galaxy. As seen in previous chapters, galaxies, like butterflies, come in a few basic shapes: disk-shaped and spheroidal systems that define rather continuous sequences of silhouettes. The challenge has been to build a meaningful and accepted sequence of forms. The story of establishing a significant classification system is instructive and fascinating to reconstruct. Atlases of galaxies will help us (Chapter 10).

Galaxy atlases, being packed with images, are essential and advantageous vehicles of any galaxy classification scheme. Using representative cases, astronomers set themselves the task of designing and producing these atlases, of which about 18 in total emerged from various inspirations; they followed their own convoluted paths, including controversial outcomes in the ways galaxies are viewed. The atlases have been a significant part of the story of understanding galaxies by classifying them.

Galaxy atlases are reference visual displays (Fig. 0.6); thus the creation of an atlas is an epistemological undertaking. Atlases are also heuristic tools for identifying and selecting

---

[4] H. Shapley and A. Ames, A Survey of the External Galaxies Brighter Than the Thirteenth Magnitude, *Annals of the Astronomical Observatory of Harvard College*, 1932, Vol. 88, pp. 41–76.

archetypes, for discovering hints of more complexity, and for making use of qualitative, then quantitative, patterns in a reproducible way. As Lorraine Daston stated, discerningly, atlases train the practitioner how to see scientific objects in unison. "The schooling of the senses is probably not qualitatively different from that undergone by the fledgling musician, cook, or weaver – as Aristotle noted, the paths to skill, on the one hand, and to understanding, on the other, pass through the same stations of perception, memory, and experience. But the scientific path is greatly straightened by the demands of collective empiricism, which require a degree of coordination seldom achieved (or desired) in the traditional arts and crafts."[5]

Starting with Flemish cartographer and mathematician Gerardus Mercator (1512–1594), the word "atlas" came to refer to large sized paper. "But it was in the nineteenth century that the oversized sheets lent their name to a much wider species of literature, the genre of systematic presentation of scientific pictures. There were atlases of every sort. From the pathology of the human body to African plants, from atlases of fossils to ones of micrographs and X rays."[6] Among books and tools of learning, scientific atlases are indeed pivotal; they are specialized products that must meet the demands of insistent communities. There is even a market for atlases, and some have even talked about their "economy."

Atlases are different from guidebooks and encyclopedia. A guidebook is a companion tool to recognize natural objects while an encyclopedia treats the whole knowledge of a field in a comprehensive manner, giving broader coverage of topics. The atlas is more systematic than a field guide and does not aim to comprehensively cover the whole discipline. To prepare for an in-depth discussion of how atlases of galaxies were put together (Chapter 10), I want to examine first how the classification of galaxies was addressed.

The Purpose of Atlases: Shaping Galaxies

The current classification scheme of galaxy shapes emerged in the first three decades of the twentieth century. To see objects "in unison" requires some organizational scheme. The classification of objects is one of the driving forces for creating scientific atlases; in reverse, the atlas images highlight and illustrate the classification scheme. A small number of shared properties are used to sort the objects into categories. As emphasized by Steven J. Dick, a "class of objects" in astronomy is different from a "species" in biology, and is much less distinctive than elements in chemistry or elementary particles in high-energy physics. While there are no transitions between elements and particles, there are intermediate states in objects of biology and astronomy. A main-sequence star, like the Sun, will transition to be a red giant, later a planetary nebula, and finally a white dwarf. An electron will not become a proton, nor will hafnium transform into tungsten; it is that or something else. "Thus, astronomical classes retain some of the ambiguities of biological species, but not

---

[5] L. Daston, On Scientific Observations, *Isis*, 2008, Vol. 99, No. 1, p. 107.
[6] P. Galison, *Image & Logic, A Material Culture of Microphysics*, Chicago: University of Chicago Press, 1997, p. 131.

their evolutionary mechanisms, introducing both practical and theoretical problems with classification."[7]

Classification is a sort of epistemological art. As beautifully stated by Gérard de Vaucouleurs, "one of the first tasks which confronts the student of almost any category of objects – atoms, molecules, plants, animals, stars, galaxies, etc. – is to arrange them in some classification scheme bringing in order and logic where nature offers only a bewildering variety of individuals."[8] American astronomer Allan Sandage (1926–2010) put forward some useful considerations. To have a "good" classification system, "some type of significant ordering is necessary; ideally it will disclose the underlying true connections among characteristics of the manifold objects."[9] Sandage then questioned what we mean by "significant" and by "true," then goes on to devote a page and a half to how the "nearly un-surmountable challenge" can be met. For galaxies, the shared property at the root of the classification scheme, that has been so far most favored, has been and continues to be *morphology* or the three-dimensional shape of a galaxy. For Sandage, de Vaucouleurs and other classifiers, morphology reveals "true connections." More recently, supplementary classification criteria using dynamic properties (e.g. counter-rotation) have been proposed.

As we saw in Chapter 5, the main shapes of non-galactic "nebulae" were recognized relatively early: first, spirals by William Parsons in 1845, then, ellipticals by Stephen Alexander (1806–1883) in 1852,[10] and much later, Magellanic or irregular systems by Knut Lundmark in 1922. The speculative work of Alexander had a striking dimension. A supporter of the Nebular Hypothesis, Alexander viewed most nebulae as systems in disintegration, which followed, rather than preceded, stellar formation (see Fig. 2.11). Attempting a grand ordering scheme in 1919, Heber Curtis threw galactic and non-galactic "nebulae" in the same bag, not unlike the scheme proposed by Max Wolf in 1908 (Chapter 4; Fig. 4.4). Curtis divided all nebulae into three groups: planetaries, diffuse nebulae in the galactic plane, and spirals in the "extragalactic" region. The distinction between "in the plane" and "out of the plane" (of the Milky Way) stuck for some time. When using the term "extragalactic nebulae," Hubble carefully made this distinction, without at first committing fully to the notion that spirals were external. However, Curtis clearly considered spirals to be a special group, systems external to the Milky Way, and the most abundantly observed, in the order of millions. Until the adoption of Hubble's orderly scheme, classification exercises were more like shots in the dark.

Along Came Edwin Hubble

The merit of Hubble resides not in his invention of the fundamental categories but the fact that he saw order enough to be able to sequence galaxy shapes in "significant" ways. Hubble

---

[7] S. J. Dick, *Discovery and Classification in Astronomy, Controversy and Consensus*, Cambridge: Cambridge University Press, 2013, p. 239.

[8] G. de Vaucouleurs, Classifying Galaxies, *Astronomical Society of the Pacific Leaflets*, 1957, No. 341, p. 329.

[9] A. R. Sandage, *Centennial History of the Carnegie Institution of Washington, Volume 1: The Mount Wilson Observatory*, Cambridge: Cambridge University Press, 2004, p. 230.

[10] S. Alexander, On the Origin of the Forms and the Present Condition of Some of the Clusters of Stars, and Several of the Nebulae, *The Astronomical Journal*, 1852, Vol. 2, pp. 95–160. The work appeared in a series of articles running through several issues.

brilliantly picked out and highlighted meaningful variations among the non-galactic "nebulae." In this way, it can be said that Hubble's "tuning fork" diagram for galaxy classification clarified and merged earlier schemes (see Plate 9.1). More systematic and thorough than its predecessors, Hubble's scheme has remained the paradigm of galaxy classification, based on morphology. It has survived important modifications and refinements later brought to it, most notably by the French astronomer Gérard de Vaucouleurs (1918–1995).

Hubble's dichotomous diagram was a powerful tool for distinguishing the categories of galaxies. Disk galaxies differ from ellipticals by their distinct three-dimensional shape and overall dynamics. For both classes, the degree of flattening and light concentration has a physical meaning and possibly relates to events that happened at the time of formation (Chapter 6). Among disk systems, bar spirals are distinguishable from normal spirals by features that indicate differences in large-scale stellar and gas kinematics. Hubble envisaged – and he was right – morphology as a tracer of fundamental processes underlying the formation, assembly and evolution of galaxies. Putting aside his usual circumspection, Hubble went even further. Influenced by the ideas of the British astrophysicist James Jeans (1877–1946) on nebular evolution, he thought that his "sequence" of shapes was evolutionary: in his view, galaxy systems started as ellipticals; with time, they became flattened disks. However, he wisely refrained from making this shaky (and incorrect) theoretical connection when he proposed his classification scheme.[11]

The great merit of Hubble's scheme was its relative simplicity and the ease with which it can be illustrated with spectacular images. For Hubble, as for the German biologist and naturalist Ernst Haeckel, "what appeared to be so complicated needed to be made plain – morphology, form were the tools."[12] This inspired the taxonomists of galaxies and led to the production of practical and splendid atlases. *The Hubble Atlas of Galaxies*, first produced by Allan Sandage in 1961, was a masterpiece that set the standard and tone for followers (Chapter 10).

In summary, driving forces for galaxy atlases were to reflect contemporary knowledge about the world of galaxies and to educate students and researchers new to the field. Furthermore, by adopting Hubble's relatively simple scheme, authors of galaxy atlases could serve a wide set of purposes as we will see in Chapter 11. Atlases also provide working objects for research programs. But they do more than that: they help to build communities of practitioners. Daston discussed extensively a great classic of functional scientific atlases: the *International Cloud Atlas* – the 1896 trilingual version with text in English, French and German – produced for meteorologists of the world to see clouds "in unison."[13] The first edition featured several printed colour plates, instead of hand-coloured plates, and even included an unusual colour photograph of a cirrus cloud. This century-old scholarly work is famous for its many successive editions and for being issued in several languages. Like this cloud atlas and atlases of other scientific disciplines, galaxy atlases aimed to assemble "standardized objects" for researchers and students. Carl Størmer's *Photographic Atlas of*

---

[11] See discussion in R. Smith, *The Expanding Universe, Astronomy's 'Great Debate' 1900–1931*, Cambridge: Cambridge University Press, 1982, pp. 147–151.

[12] O. Breidbach, *Visions of Nature: The Art and Science of Ernst Haeckel*, Munich: Prestel, 2006, p. 274.

[13] L. Daston, On Scientific Observations, *Isis*, 2008, Vol. 99, No. 1, pp. 97–100.

Auroral Forms attempted the same.[14] Atlases are like sorts of bibles that bring practitioners to a common "creed." On the practical side, they are the handbooks that guide day-to-day work on galaxies.

Through its production and usage, a scientific atlas represents the epitome of scientific observation (see Introduction), building communities by observing together. Lorraine Daston and Elizabeth Lunbeck have brilliantly described this process as collective empiricism. It is "the collection, transmission, and distillation of the experience of many inquirers into weather proverbs, astronomical tables, medical regimens, botanical descriptions, socioeconomic statistics, and a myriad of other findings."[15] The atlas became the tool to "calibrate the eyes and hands" of the dispersed community of galaxy observers and students (more on this in the next two chapters).

Classification for a Purpose

All atlases use and highlight a classification system of the scientific objects of interest. They also do more. They try to convey lots of information: for galaxies, images can be supported by quantitative photometric measurements in order to disentangle the mix of structures. The photometric maps pinpoint the different kinematic or evolutionary components of galaxies. Taken at different orientation angles, radial light profiles help to separate the central bright nuclei and amorphous bulges from the flatter and more extended exponential disks; and they provide more specific details, such as the spatial enhancements of star concentration, or bars/ovals/rings that highlight dynamic instabilities. Finally, the strength of dark lanes gives a sense of the amount and distribution of dust in a galaxy. To the perspicuous eye and trained user, atlas images illustrate the effect of internal perturbations (bars), secular evolution, gas accretion and nuclear activity. By organizing all of this, a classification scheme helps to interpret morphology as it relates to environmental density and the merging history of galaxies.

As the morphology provides clues on the formation and evolution of the gas and stars of a galaxy, products of computer simulations can be related to the observed shapes of archetypal objects. Hence, atlases, assembling sequences of observed configurations, become useful collections of working objects for theoreticians and simulators to test their models on.

Brief Historical Overview of a Meandering Process

Hubble was not the first to tackle the challenge of classifying nebulae. To grasp the success of his scheme, it is useful to recall a bit of history. The first attempts at classifying "nebulae" can be seen in the 1811 drawings of William Herschel (Fig. 2.4) and, more spectacularly, in the publications of works conducted later under the leadership of William Parsons, the Third

---

[14] C. Størmer, *Photographic Atlas of Auroral Forms*, Oslo: International Geodetic and Geophysical Union, 1930.
[15] L. Daston and E. Lunbeck, Observing Together: Communities, in *Histories of Scientific Observation*, Chicago: University of Chicago Press, 2011, p. 369.

Earl of Rosse, at Birr Castle, Ireland, where a spiral structure was found in several "nebulae" (Fig. 0.2). As we saw in Chapter 2, Parsons encouraged the faithful visual rendition of "nebulae" through drawings made by careful observers using the large Birr Castle reflectors over three decades (1848–1878).[16,17] The Birr Castle observers found spirality in dozens of other "nebulae." Although some cases turned out to be false identifications, most of these objects were later confirmed by photography to be spiral galaxies.[18,19] The nineteenth-century drawings captured the essential shape and important details of the galaxy structures and of the satellite galaxies surprisingly well.

In a rather long-winded article in 1852, the American astronomer and mathematician Stephen Alexander tried to make a grand synthesis of all nebular forms and star clusters: "spheroid," "elliptic," "annular," "spiral," "globular" (see Fig. 2.11).[20] He illustrated all his examples with drawings from Birr Castle and others produced by John Herschel. He tried to organize "nebulae" by their shape and even predicted hypothetical shapes. The main merit of Alexander's highly speculative article was the recognition of the spheroidal/ellipsoidal shapes of non-galactic nebulae.

The British amateur astronomers Andrew Ainslie Common and Isaac Roberts used photography and their own observations based on telescopes with improved designs and glass mirrors (Fig. 3.1). The advent of sophisticated telescope mounts as well as automatic drives revolutionized the observation of faint objects, in particular the "nebulae." The use of negative prints was favored: a negative print reproduced the details with a greater fidelity and was better able to handle the wide contrast between the very bright nucleus (generally saturated), the bright bulge and the faint outer arms or inter-arm regions. Roberts published two spectacular books of his photographs of "nebulae" and star clusters.[21] They did not attempt to classify, nor were they atlases in the scholarly sense, but they defined the way forward.

It was photography, not drawing, that became the obvious means for identifying categories. Max Wolf used photographic material to establish an early system of classes of "nebulae" (Chapter 4). He did not differentiate between galactic and non-galactic nebulae when proposing 23 classes distinguished by alphabetical letters a to w (with a h_o class). Today, classes o to w can be recognized as being spirals with different inclinations and strength of dust lanes, while a to n appear to be a mix of planetary nebulae and non-galactic spheroidal systems. However, Wolf had no idea of the nature of the objects he was trying to organize. His scheme remained a temporary "filler" and of marginal interest until something better came along.

---

[16] Third Earl of Rosse, On the Construction of Specula of Six-Feet Aperture: and a Selection from the Observations of Nebulae Made with Them, *Philosophical Transactions of the Royal Society of London*, 1861, Vol. 151, pp. 681–745.

[17] Fourth Earl of Rosse, Observations of Nebulae and Clusters of Stars made with the Six-foot and Three-foot Reflectors at Birr Castle, from the year 1848 up to about the year 1878, *The Scientific Transactions of the Royal Dublin Society*, 1879, Vol. 11.

[18] D. W. Dewhirst and M. Hoskin, The Rosse Spirals, *Journal for the History of Astronomy*, 1991, Vol. XXII, pp. 257–266.

[19] Ronald Buta did an analysis of the Birr Castle observations in 2010 and concluded that the Birr observers saw spiral structure in 75 bright galaxies.

[20] S. Alexander, On the Origin of the Forms and the Present Condition of Some of the Clusters of Stars, and Several of the Nebulae, *The Astronomical Journal*, 1852, Vol. 2, pp. 95–160.

[21] I. Roberts, *A Selection of Photographs of Stars, Star-clusters and Nebulae*, Vol. I, London: The Universal Press, 1893; I. Roberts, *A Selection of Photographs of Stars, Star-clusters and Nebulae, Volume II*, London: "Knowledge" Office, 1899.

Then, being in possession of photographs of nearly 500 spirals, Heber Doust Curtis presented the first major compendium of photographic galaxy images in 1918, with a special focus on spirals.[22,23] The Lick astronomer even thought that all non-galactic "nebulae" were spirals, predicting – wrongly – that ellipticals would be seen as spirals with higher angular resolution. Curtis illustrated the shape of the spirals with attention to the dust, and correctly inferred that it affected the appearance of the objects. Plates of dozens of spirals photographed with the 36-inch Crossley telescope were presented as a mini atlas at the end of the long article. The images were displayed as sequences of shapes to illustrate the relative importance and spatial distribution of internal dust.

A Contentious Path to Galaxy Classification

John Henry Reynolds (1874–1949) was a highly successful British industrialist – another astronomy dilettante businessman and gentlemanly scientist – who was passionate about astronomy.[24] He had constructed his own 28-inch telescope and he served as the president of the Royal Astronomical Society. Unusually, Reynolds played in the same field as professionals. He published his own classification of spiral galaxies (calling them "spindle" nebulae) in 1920, well ahead of Hubble, who corresponded with him.[25] On the basis of the degree of central condensation and granularity of the "outer whorls" or structure of the circular pattern, Reynolds had created seven classes of spirals some of which turned out to be identical to Hubble's. Hubble referred much later and only in passing to Reynolds' works, and not to the original work that must certainly have inspired him.[26]

It is well known that Hubble was extremely sensitive regarding the priority of discovery and of the proposal of concepts. He did not easily accept nor recognize competition. This is clearly highlighted by the ensuing Hubble–Lundmark controversy that showed a trait of Hubble's character.[27] This confrontation was also a sign that galaxy classification had become an intense battlefield of clashing concepts – and personalities. Hubble came up with his famous "tuning fork" diagram in the mid 1920s (Plate 9.1).[28] However, as early as 1920, the Swedish astronomer Knut Lundmark (1889–1958) had proposed a classification system of "anagalactic" nebulae that was quite analogous to Hubble's scheme (Fig. 9.2).[29] Lundmark's scheme contained some slight but important differences,

[22] H. D. Curtis, Descriptions of 762 Nebulae and Clusters Photographed with the Crossley Reflector, *Publications of the Lick Observatory*, 1918, Vol. XIII, Part I, pp. 9–42.

[23] H. D. Curtis, A Study of Occulting Effects in the Spirals, *Publications of the Lick Observatory*, 1918, Vol. XIII, Part II, pp. 43–55.

[24] A fine review and assessment of John H. Reynolds' work and contributions are presented in D. L. Block and K. Freeman, *Shrouds of the Night, Masks of the Milky Way and Our Awesome New View of Galaxies*, New York: Springer, 2008, pp. 183–213.

[25] John H. Reynolds, Photometric Measure of the Nuclei of some Typical Spiral Nebulae, *Monthly Notices of the Royal Astronomical Society*, 1920, Vol. 80, pp. 746–753.

[26] D. L. Block and K. Freeman, op. cit., 2008, pp. 200–203.

[27] This episode has been particularly well discussed in Part III, Chapter 6 of R. Berendzen, R. Hart and D. Seely, *Man Discovers the Galaxies*, New York: Science History Publications, 1976.

[28] E. P. Hubble, Extra-Galactic Nebulae, *Contributions from the Mount Wilson Observatory*, Carnegie Institution of Washington, 1926, Vol. 324, pp. 1–49; Extra-galactic Nebulae, *The Astrophysical Journal*, 1927, Vol. 64, pp. 321–369.

[29] Anagalactic was a word equivalent to non-galactic used mainly by European astronomers. Galaxies was the word used almost immediately after Hubble's death. Hubble stuck to the end to "extra-galactic nebulae."

Fig. 9.2 Knut Lundmark as a student in 1908. Credit: unknown photographer at H. Tegström & Co.

especially in emphasizing the degree of light concentration towards the center. In 1926, a furious Hubble wrote to Lundmark, accusing him of plagiarism and of having stolen his classification scheme following the 1925 Cambridge Astronomical International Union (IAU) meeting, where Hubble had presented his proposal.

As shown by the Finnish astronomer Pekka Teerikorpi, Hubble had incorrectly and unfairly accused Lundmark.[30] Teerikorpi demonstrates that Lundmark had worked on galaxy classification some years before his 1926 paper; the concepts of his system and the sequence of the main classes had appeared in print before the 1925 Cambridge IAU meeting. Finally, "Lundmark's and Hubble's systems were in fact so different that they hardly could have served as models for each other's." Allan Sandage also thought Hubble had overreacted and treated Lundmark unfairly. Sandage wrote that the "exchanges poisoned the association between Hubble and Lundmark for years. It would not be mended until the

[30] P. Teerikorpi, Lundmark's Unpublished 1922 Nebula Classification, *Journal for the History of Astronomy*, 1989, Vol. 20, pp. 165–170.

1950s, when Hubble invited Lundmark to inspect the archival plate file of galaxies that had been photographed by the new 200-inch Palomar reflector."[31]

This clash certainly illustrates how possessive scientists can be of their ideas, but it highlights the importance galaxies were now taking among professionals as a rapidly developing research field. Also, the Lundmark–Hubble skirmish shows that a good concept may burgeon simultaneously in several minds which are trying to tackle the same problem. For example, James Jeans had come up with a "Y" classification to emphasize that the "great nebulae" could not all be placed in a continuous sequence.[32] Some say that the Hubble tuning fork could be viewed as just an extended Y scheme. "In popular astronomical textbooks, we only read of the Hubble classification classes, the Hubble tuning fork and the Hubble luminosity profile for elliptical galaxies. Behind stage loomed the giant Mr. Reynolds, a man whose name is almost unknown to students of astronomy."[33] There goes the whiggish history of science. Mark Twain had already denounced Stigler's law of eponymy in harsher words: "It's just the way, in this world. One person does the thing, and the other one gets the monument."

The Hubble Sequence

Let us come back to the Hubble system of galaxy classification, often called the "Hubble sequence" (Plate 9.1). Although Hubble himself would have liked it to be so, the "sequence" does not represent an evolutionary trend. Hubble divided galaxies into three broad classes, following their appearance at visible wavelengths: ellipticals, lenticulars and spirals. He added a fourth class, the irregulars (or Lundmark's "Magellanic"), characterized by a lack of symmetry. Indeed, many galaxies could not be classified; several of these unclassifiable cases gave rise to debate and controversy and were later recognized as "peculiar."

On the Hubble sequence, spirals divide into two categories, normal (e.g. Messier 81, Plate 7.1) and barred (e.g. NGC 7424, Plate 6.4), giving two branches, hence, "tuning fork." Hubble's tuning-fork diagram runs from the left with the ellipticals, with the lenticulars at the center, and to the right, the two branches of the fork of the normal spirals (top branch) and barred spirals (bottom branch). The irregulars (e.g. NGC 4449, Plate 6.2) are hanging at the right for apparent lack of morphological connection with the rest. Hubble nicely presented his perspectives and the purposes of his scheme in a fine little book, *The Realm of Nebulae*. The 1936 book is a classic and remains a most useful source in the context of historical development.[34]

Let me add a few technical details on the Hubble classification that will be important for the discussion of atlases of galaxies in the next chapter. The symbol E, for elliptical,

---

[31] A. R. Sandage, *Centennial History of the Carnegie Institution of Washington, Volume 1: The Mount Wilson Observatory*, Cambridge: Cambridge University Press, 2004, p. 488.
[32] J. Jeans, *Astronomy and Cosmogony*, Cambridge: Cambridge University Press, 1928, Figure 53.
[33] D. L. Block and K. Freeman, *Shrouds of the Night, Masks of the Milky Way and Our Awesome New View of Galaxies*, New York: Springer, 2008, p. 212, citing Mark Twain from *Tom Sawyer's Conspiracy* (published posthumously).
[34] E. P. Hubble, *The Realm of Nebulae*, New Haven: Yale University Press, 1936.

is followed by an integer (0 for spheroidal to 7 for the most flattened ellipsoid), which indicates the degree of ellipticity multiplied by 10; it is defined the normal way for an ellipse as $e = 1 - (b/a)$, a and b being the semi-major and semi-minor axis; E is $10 \times e$. Ellipticals are truly ellipsoidal structures in three-dimensional space. At the center of the fork are the lenticulars denoted by the symbol S0. They have a bright, large, central bulge, very similar to elliptical systems, and are surrounded by a disk-like structure as in spirals. Lenticulars, although recognized as a separate class, were not understood in Hubble's time, and many successive researchers scratched their heads over them until they fell into place quite recently (Chapter 11). Lenticulars look like the smooth spindle-shaped galaxies seen at the top and bottom edge of the field on Plate 6.3.

Spirals display the most spectacular shapes. They are multi-component systems, dominated by a flattened disk, with gas and star-forming regions concentrating generally in a two-armed spiral structure. Spirals also have a bright central region, the bulge. Harlow Shapley and John S. Paraskevopoulos subdivided the Sc end of the spiral sequence into Sc and Sd to describe systems with greater arm openness and a lack of central condensation. Thus, spirals are denoted by the symbol S followed by a, b, c or d as suffixes.[35] Those with bigger bulges and smoother arms are the Sa spirals, while Sd spirals show loose fragmentary arms and have no bulge. The suffixes apply to barred spirals as well, and are then denoted as SBa, ..., SBd.

Because of their lack of bulge or of structural symmetry, irregular galaxies do not fit into the Hubble sequence. Hubble had a hard time with these intractable entities. Gérard de Vaucouleurs, who later refined Hubble's scheme, managed to integrate them into his global scheme in a less arbitrary fashion. De Vaucouleurs divided irregulars into those with some faint spiral structure as Irr I (e.g. the Large Magellanic Cloud, Plate 6.5) and irregulars with smooth features as Irr II; those with no obvious structure were described as Im (e.g. the Small Magellanic Cloud). Hence, de Vaucouleurs recognized clearly the category invented by Lundmark in 1922. Another Swedish astronomer, Erik Holmberg (1908–2000), introduced finer divisions.[36] All this fine-tuning of viewing galaxy shapes led to the entire classes for spirals used today: Sa, Sab, Sb, Sbc, Sc, Scd, Sd, Sdm, Sm and Im, the latter two corresponding to Lundmark's Magellanic types. Finally, ellipticals and lenticulars are sometimes referred to as "early type" while spirals and irregulars are "late type." This unfortunate nomenclature has no meaning in terms of temporal evolution. It is very confusing and should be avoided.

The modern classification system of galaxies is not without parallel to the binomial nomenclature for naming living species introduced by Carl Linnaeus in 1753.[37] Linnaeus used a binomial name, where the first part identifies the genus of the species and the second part identifies the species within the genus. Galaxy shapes are related in much simpler ways

[35] H. Shapley and J. S. Paraskevopoulos, Galactic and Extragalactic Studies, III: Photographs of Thirty Southern Nebulae and Clusters, *Proceedings of the National Academy of Sciences of the United States*, 1940, Vol. 26, pp. 31–36.

[36] E. Holmberg, A Photometric Study of Nearby Galaxies, *Meddelanden fran Lunds Astronomiska Observatorium*, 1950, Ser. II, 128, 1.

[37] C. Linnaeus, *Species Plantarum*, Stockholm: Holmiae, Impensis Laurentii Salvii, 1753.

than living species, but the galaxy classes Sa, SBc, S0 or E1, have an analogous binomial structure that is a powerful visual descriptor. Praising Hubble's scheme, Allan Sandage has summarized it well: "The great merit of Hubble's system is the bin size of the classification boxes."[38] Big enough to be significantly inclusive and narrow enough that even the neophyte can apply the scheme successfully. However, as we will see later, Gérard de Vaucouleurs found the bin sizes of Hubble's original system too broad, and applied a finer division.

Defining a Systematic Sample

A useful atlas depends on a well-defined and agreed sample of objects. And the sample must be large enough. The Harvard astronomers Shapley and Ames published "A Survey of the External Galaxies Brighter than the Thirteenth Magnitude" in 1932.[39] This catalogue of 1,249 bright galaxies was based on a photographic survey conducted from Arequipa, Peru, using two tiny telescopes, the 2-inch Ross–Tessar and 2-inch Zeiss–Tessar. Small lenses indeed, but they served two clear purposes: the small scale of the photographs gave the advantage that most nebular objects looked sufficiently stellar for reliable brightness intercomparisons and precise apparent magnitude determinations.

With their photometric catalogue, Shapley and Ames aimed "to provide a systematic census of the inner parts of the metagalactic system where heretofore no comprehensive photometry has been available." Although it had no images, the Shapley–Ames survey may be considered as the seed for most galaxy atlases that were to be produced in the following 80 years. It provided the basic sample for the study of "the uniformity of distribution of the galaxies, the clustering of the nearer systems, the relation of the apparent distribution to the obscuring clouds in low galactic latitude, and similar problems."

The Shapley–Ames catalogue included a discussion on the sizes and shapes of galaxies "having Hubble as the authority for the description." The catalogue also gave an overview of several early schemes of galaxy classification (those of Wolf, Curtis, Reynolds, Reinmuth and Hubble) and of their cross-correspondence. Karl Reinmuth had adopted Wolf's scheme (Fig. 4.4).[40] Catalogues are essential to maintaining the usage of a classification system. They provide systematic lists of objects as complete sets or representative ones. This certainly worked most fruitfully for the Shapley–Ames catalogue.

The Shapley–Ames catalogue sample was indeed the basis of several later works, in particular the long-term program conducted by Allan Sandage and collaborators at the Carnegie Observatories. The later revisions of the catalogue (*The Revised Shapley–Ames Catalog of Bright Galaxies*, or *RSA*) by Sandage and Tammann (1981, 1987) were solidly

---

[38] A. R. Sandage, *Centennial History of the Carnegie Institution of Washington, Volume 1: The Mount Wilson Observatory*, Cambridge: Cambridge University Press, 2004, p. 489.

[39] H. Shapley and A. Ames, A Survey of the External Galaxies Brighter than the Thirteenth Magnitude, *Annals of the Astronomical Observatory of Harvard College*, 1932, Vol. 88, pp. 41–76.

[40] K. Reinmuth, Die Herschel-Nebel nach Aufnahmen der Künigstuhl-Sternwarte, *Veroeffentlichungen der Badischen Sternwarte zu Heidelberg*, Berlin and Leipzig: Walter de Gruyter, 1926, Vol. VI. Reinmuth used Wolf types, and distinguished the directions of spiral arms.

anchored to the work of Harlow Shapley and Adelaide Ames. The most spectacular outcome was the massive *The Carnegie Atlas of Galaxies* by Sandage and Bedke in 1994 that will be discussed in Chapter 10.[41] Allan Sandage wrote repeatedly about galaxy classification, its development and the role of the *RSA* based on the Shapley–Ames catalogue.[42] Several other galaxy surveys used the Shapley–Ames catalogue to define and select samples of galaxies.

Gérard de Vaucouleurs "Improvements"

In parallel to Sandage's work on galaxy classification, Gérard de Vaucouleurs published a pioneering work on the classification and morphology of galaxies in 1959.[43] Entering the stage from the side, de Vaucouleurs thoroughly reconsidered the history of the morphological classification of galaxies.

De Vaucouleurs had spent the period from 1951 to 1957 at Mount Stromlo Observatory outside Canberra, Australia (see Fig. 10.1), then named the Commonwealth Observatory. While there, he carried out a survey of the bright galaxies listed in the Shapley–Ames catalogue south of the declination –30°. He had been using the 30-inch Reynolds reflector, which had been donated to the Australian institution by John Reynolds, the same individual whom we saw proposing the classification scheme that inspired Hubble. It was the second largest in the southern hemisphere until the 1950s.[44] In 1957, de Vaucouleurs published his survey in a finely illustrated article that was a vanguard of the future atlases of galaxies.[45] Plate I of the paper is an initial two-dimensional representation of de Vaucouleurs' modified scheme (Fig. 9.3). Later, de Vaucouleurs changed this evocative visual representation into a three-dimensional one. He also introduced a more detailed notation system to better distinguish shapes of galaxies and their finer structural types. The improved classification scheme was a complete and most elegant classification system based on morphology, but was perceived as more complicated than Hubble's.

While de Vaucouleurs' articles were not atlases, the numerous plates at the end of his articles were a carefully assembled set of images that illustrated most extensively his proposed classification scheme. As such, they were sorts of mini-atlases. Like Hubble and Sandage, de Vaucouleurs sorted galaxies along a sequence, with ellipticals starting on the left, and going to the right as shapes flattened; then the spirals and disk systems taper, with the irregulars at the extreme right, in a systematized way. His most original contribution was to add a third dimension to Hubble's diagram (Fig. 9.4). This shrunk the broad divisions of Hubble and allowed finer bins for sorting objects. Set orthogonal to the sequence,

---

[41] A. R. Sandage and J. Bedke, *The Carnegie Atlas of Galaxies*, Washington DC: Carnegie Institution of Washington Publications, 1994.

[42] See for example A. Sandage, Classification and Stellar Content of Galaxies Obtained from Direct Photography, in *Galaxies and the Universe*, A. Sandage, M. Sandage and J. Kristian (editors), Chicago: University of Chicago Press, pp. 1–55.

[43] G. de Vaucouleurs, Classification and Morphology of External Galaxies, *Handbuch der Physik*, 1959, Vol. 59, pp. 275–310.

[44] The Reynolds telescope was destroyed during the 2003 Australian firestorm. The "Melbourne Telescope" was the largest telescope in the south from 1869, though the design of its tube and mounting prevented it from being a truly useful instrument until its extensive modification in the early 1960s. It, too, was destroyed in the 2003 firestorm.

[45] G. de Vaucouleurs, Survey of Bright Galaxies south of –35° Declination, with the 30-inch Reynolds Reflector (1952–1955), *Memoirs of the Commonwealth Observatory*, 1956, Vol. III, No. 13, + 8 plates.

PLATE I.

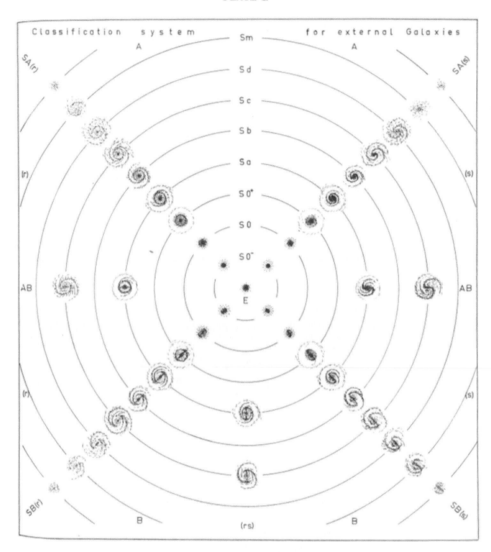

Fig. 9.3 "Mark I" version of de Vaucouleurs' classification system of galaxies proposed in 1957. From G. de Vaucouleurs (1957), *Memoirs of the Commonwealth Observatory*.

planes appear like the faces of a clock and convey secondary morphological features such as bars, rings and lenses, with a gradation like on a clock (Fig. 9.5).[46] Visually, it is a most elegant arrangement.

---

[46] This effort by de Vaucouleurs led to another massive catalogue of bright galaxies (*RC3*) that has been used extensively by researchers: G. de Vaucouleurs, et al., *Third Reference Catalogue of Bright Galaxies*, Three volumes, New York: Springer-Verlag, 1991. It contains more than 23,000 objects with extensive size, photometric and kinematic data.

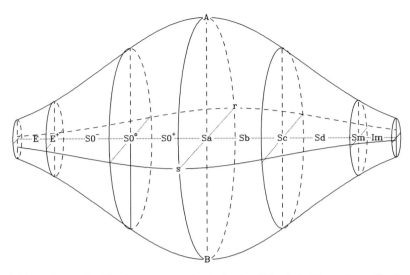

Fig. 9.4 The volume classification system of galaxies by Gérard de Vaucouleurs. Credit: G. de Vaucouleurs (1959), *Handbuch der Physik*.

Gérard de Vaucouleurs' main contributions to galaxy classification were:

- Introducing the SA for intermediate bar (SAB) classifications between the pure spiral S and the strongly barred SB.
- Recognizing "outer pseudo-rings" as an important aspect of morphology.
- Strongly recognizing the apparent continuity in galaxy structure, to the point that de Vaucouleurs felt he could place any given galaxy at a very specific point within his classification volume.
- Defining a consistent way of classifying lenticulars, or S0 galaxies, which makes the category a reasonable stage and not a "garbage bin."[47]

Variation on a Theme, a Spectral Classification for Galaxies

Although convergence on the classification criterion of morphology occurred relatively quickly, there were attempts to use other properties to distinguish and classify galaxies. The most serious endeavor was that of American astrophysicist William Wilson Morgan (1906–1994), a leader in stellar spectroscopy. With fellow spectroscopists Philip Childs Keenan (1908–2000) and Edith Kellman (1911–2007), Morgan had previously devised the very successful and widely used MKK system for classifying stellar spectra, an autonomous scheme "without having to appeal to any theoretical picture."[48] As much as stars could be

---

[47] I am indebted to R. Buta for highlighting these contributions (e-mail to author 22 February, 2015).

[48] W. W. Morgan, A Morphological Life, *Annual Review of Astronomy and Astrophysics*, 1988, p. 4. The standard MKK system was based on a strict comparison of the intensities of well-picked spectral lines. The main physical determinant of line intensity is temperature, with secondary influences of abundance and gas pressure or gravity.

Fig. 9.5 Cross-section of the de Vaucouleurs system near class Sb. From Buta et al. (2007), *The de Vaucouleurs Atlas of Galaxies*. Courtesy of Cambridge University Press.

easily classified, Morgan felt that the fantastic diversity of galaxies noted by Reynolds was truly challenging. "The classification of the optical forms of galaxies is a very different problem from that of the classification of stellar spectra. The ordered appearance of the spectrograms contrasts sharply with the apparent disorder of the forms of many galaxies, even in the case of galaxies possessing a degree of order, we find them in fantastic variety."[49] This did not stop Morgan from giving it a try.

[49] W. W. Morgan, A Morphological Life, *Annual Review of Astronomy and Astrophysics*, 1988, p. 6.

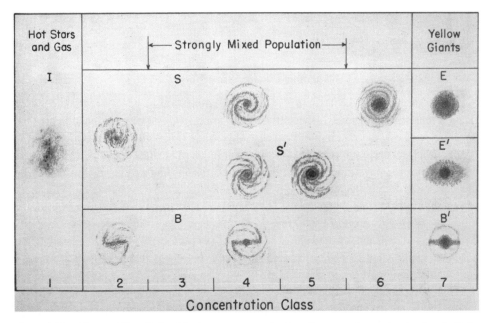

Fig. 9.6 The Morgan–Mayall classification system emphasizes the types of stars dominating the light of the galaxies. Credit: From Morgan (1962), *The Astrophysical Journal.* © AAS. Reproduced with permission.

The essence of the venture was to distinguish in galaxy spectra the types of stars that were most effective in contributing to the integrated light. Working later with the American astronomer Nicholas Ulrich Mayall (1906–1993), Morgan proposed, in 1957, a spectral classification of giant galaxies using quantifiable properties of galaxies as opposed to visual estimates employed for morphological screening.[50] He and his colleagues used the spectra of galaxies as a whole but isolated the spectrum of the nuclear region in cases where this region was dominant; there could be noticeable spectral differences between the nucleus and the surrounding regions. Using the characteristics of the galaxy spectrum in the blue spectral region and the degree of central concentration of light, Morgan and Mayall's spectroscopic indicators produced seven groups (Fig. 9.6).

Because the relative size of the amorphous central region played a role, the system mimicked one of Reynolds' and Hubble's main criteria, the central concentration of brightness. Despite its relative rigour, the Morgan–Mayall system was complicated and did not catch on with the community. Some of its elements remain in use, for example the notation of "D systems" for giant ellipticals (often associated with a strong radio source) surrounded by a huge external envelope of stars. When located close to the centers of clusters of galaxies, these giant ellipticals are called cD galaxies. Beyond this, the Morgan–Mayall spectroscopic scheme has seen little usage, probably because the integrated spectra of a galaxy

[50] W. W. Morgan, Some Characteristics of Galaxies, *The Astrophysical Journal*, 1962, Vol. 135, pp. 1–10.

can be so diverse and gives little indication of the nature of the object, especially for the neophyte or the less familiar researcher.

What does Classification Miss Out?

As will be discussed in the next chapter, many researchers raised caveats regarding classification schemes for galaxies, especially Hubble's one. The Hubble classification uses the appearance of galaxies on images. But these objects are three-dimensional. Orientation in space (and dust obscuration) creates viewing aspects or silhouettes that are not necessarily related to intrinsic physical properties. Furthermore, the appearance of a galaxy depends on the wavelength of the light that can pass through the filters used (Chapter 4). Notwithstanding these limitations, and whether the Hubble or the de Vaucouleurs scheme is used, the morphological types correlate with several key physical properties: luminosities, masses, colours, gas content and star-formation rate. Consequently, the morphological classification schemes have been useful and continue to be, despite the somewhat subjective and arbitrary steps of assigning galaxies to classes.

Another problem hampering a simple classification model is that there are many galaxies in unusual states. As de Vaucouleurs warned, "It cannot be too strongly emphasized that the regular, beautiful spiral patterns of such objects as M 31 [Messier 31], M 33, M 51, M 81, M 101, etc. represent exceptions rather than the rule."[51] As will be seen in Chapter 10, several of the peculiar galaxies appear to be "normal" galaxies (spirals, ellipticals or even irregulars) in the process of interacting or merging. Their shapes are distorted through the formation of tidal tails and their appearances are modified by enhanced star formation and dust lanes. No standard classification scheme can be applied to the "pathological" or "freak" cases, terms used by Harlow Shapley and Walter Baade (see, e.g., Plates 6.6 and 11.1). However, with the aid of computer simulations, observers and theoreticians have been able to identify the various stages of the merging process (Chapter 6).

Despite these caveats, it is generally accepted that galaxy morphology holds answers to several fundamental questions. Ronald Buta reminds us of the main questions that hinge on an understanding of galaxy formation and evolution: "Why is the Hubble sequence a continuous sequence, and what physical parameters underlie this continuity? What is the role of angular momentum, dissipation, and galaxy–galaxy interactions? How do the various patterns originate, and how do they change with time?"[52] And a more difficult but exciting question is how dark matter shapes galaxies.

Let us now review the atlases of galaxies published in the last 80 years, and the role they played in framing the collective empiricism of students of galaxies.

---

[51] G. de Vaucouleurs, Classifying Galaxies, *Astronomical Society of the Pacific Leaflets*, 1957, No. 341, p. 332.

[52] R. Buta, Galaxy Morphology and Classification, in *The World of Galaxies*, H. G. Corwin Jr. and L. Bottinelli (editors), New York: Springer-Verlag, 1989, pp. 29–44. See also R. Buta, Galaxy Morphology, in *Planets, Stars, and Stellar Systems, Volume 6*, W. C. Keel (editor), Dordrecht: Springer, 2013, pp. 1–89; R. Buta, Galaxy Morphology, in *Secular Evolution of Galaxies*, J. Falcón-Barroso and J. H. Knapen (editors), Cambridge: Cambridge University Press, 2013, pp. 155–258.

10

Atlases of Galaxies, Picturing "Island-Universes"

The human understanding, from its peculiar nature, easily supposes a greater degree of order and equality in things than it really finds.

Francis Bacon[1]

In fact, when looked at closely enough, every galaxy is peculiar.

Halton Arp[2]

Why Are There So Many Galaxy Atlases?

I was 16 or 17 years old when I received my first major scientific book as a gift from my mother; it was *The Hubble Atlas of Galaxies* by Allan Sandage. I could hardly read English then, but the black-and-white halftone images of the atlas were stunning; they have stayed imprinted in my mind. This galaxy atlas became for me, as for many other young people of the 1960s, the beginning of an amazing journey, a career in astronomy. Reading a few elementary astronomy books, and some slightly outdated texts, had already whet my appetite. It was *Larousse Encyclopedia of Astronomy* by Lucien Rudaux and Gérard de Vaucouleurs that got me hooked, a massive work that the young French astronomer de Vaucouleurs had updated and refreshed in a most professional way following the death of lead author Lucien Rudaux (1874–1947).[3]

Gérard de Vaucouleurs, who, as we saw in the previous chapter, proposed a more detailed classification scheme than Hubble's one, arrived into astronomy as a keen and passionate amateur. Young de Vaucouleurs was at first interested in the planets and asteroids of the solar system. He met his wife and astronomy co-worker while both were studying at La Sorbonne in Paris (Fig. 10.1). Antoinette (Piétra) de Vaucouleurs (1921–1987) became a key collaborator in everything that Gérard was involved with.[4] The time spent by the de

---

[1] F. Bacon, Aphorism 45, *Novum Organum*, 1620. See Hanover Historical Texts Project on-line.
[2] H. C. Arp, *The Atlas of Peculiar Galaxies*, Pasadena: California Institute of Technology, 1966.
[3] The English version was a translation and revision from the French of *Astronomie, les astres, l'univers*, Paris: Librairie Larousse, 1948.
[4] M. Capaccioli and H. Corwin Jr. (editors), *Gérard and Antoinette de Vaucouleurs: A Life for Astronomy*, Singapore: World Scientific, 1989.

Fig. 10.1 Gérard and Antoinette de Vaucouleurs in Paris, 1962. Image provided via Ken Freeman, photographer unknown.

Vaucouleurs at the Commonwealth Observatory on Mount Stromlo in Australia, from 1951 to 1957, was most productive. Throughout his career, Gérard was passionate about and most proficient in photography.[5] He wrote several books and articles about it. The couple's early works established them as key contributors and leaders of galaxy astronomy. In addition to exploring the morphology of galaxies and of the Magellanic Clouds, two important pieces of research came out from their time in Australia: (1) the relationship between mass and luminosity in elliptical galaxies, and (2) the evidence for the organization of galaxy clusters into larger structures or superclusters.

In 1958, the de Vaucouleurs went to the United States to work first at the Lowell Observatory, then Harvard College Observatory. They also visited the Mount Wilson Observatory headquarters in Pasadena, California, as guests of Allan Sandage. Sandage gave Gérard access to the Mount Wilson and Palomar observatories' archives of superb photographic

---

[5] D. F. Malin, Astronomical Photography, in *Gérard and Antoinette de Vaucouleurs: A Life for Astronomy*, M. Capaccioli and H. Corwin Jr. (editors), Singapore: World Scientific, 1989, pp. 53–64.

plates, and encouraged Gérard to develop his own classification scheme.[6] At that point in his career, Sandage had obtained a relatively small percentage of the total number of plates taken at these observatories. Then the de Vaucouleurs moved to the University of Texas and McDonald Observatory in 1960 where they spent the rest of their careers. The couple traveled together all the time. Wherever they were visiting, they always requested a quiet office where they would work. They devoted their lives to astronomy, and spent little time away from it. Antoinette was a keen observer herself. In 1957, she had noted that the brightness of the central regions of several Seyfert galaxies (Chapter 7) varied perceptibly over a period as short as a month. A skeptical Gérard told Antoinette at that time that if anything in the universe did not vary, it was surely the galaxies. "That was the greatest error of my life!" Gérard said.[7] A rapid variability of the nuclear luminosity of Seyfert galaxies is now commonly observed and understood as energetic events taking place in the accretion disks around central black holes. Antoinette passed away in 1987. Gérard de Vaucouleurs, who died in 1995, is considered a giant of twentieth-century astronomy, and Antoinette was a pillar of the modern extragalactic field of astronomy.

The Ingredients of Galaxy Atlases

As seen in the previous chapter, the hundreds of images of galaxies that had been obtained, especially in the first two decades of the twentieth century, provided a solid basis for artisans of galaxy classification to develop discriminating and meaningful schemes. Despite some limitations (for example, no two galaxies look exactly the same), the shapes of galaxies offer a reliable, relatively non-arbitrary, epistemological basis for constructing atlases. Sorting galaxies by morphology turned out to be possible, and meaningful, as several properties correlate with shape. Many had proposed that morphology can be traced to fundamental processes governing the formation, assembly and evolution of galaxies. By 1960, this audacious and relatively simple idea, which was put forward by several during the nineteenth century, had been vindicated.

Well-established and working at the world's foremost observatory, Edwin Hubble ascertained his overview of galaxy shapes concisely: "About 3 per cent are irregulars, but the remaining nebulae fall into a sequence of type forms characterized by rotational symmetry about dominating nuclei. The sequence is composed of two sections, the elliptical nebulae and the spirals, which merge into each other." In his 1936 *The Realm of Nebulae*, Hubble published several fine sets of galaxy photographs of different morphology to illustrate his proposed classification scheme.[8] Although sometimes accused of ignoring others' merits and efforts to tackle the tricky subject, it is fair to say that Edwin Hubble has been a prime mover in the contemporary classification of galaxies.

---

[6] E. M. Burbidge, *Gérard de Vaucouleurs 1918–1995, A Biographical Memoir*, Washington: National Academy Press, 2002, Vol. 82, pp. 1–17.
[7] E. M. Burbidge, *Gérard de Vaucouleurs 1918–1995, A Biographical Memoir*, Washington: National Academy Press, 2002, Vol. 82, p. 11.
[8] E. P. Hubble, *The Realm of Nebulae*, New Haven: Yale University Press, 1936.

Hubble's original scheme was revised and expanded by de Vaucouleurs, in 1959,[9] and by Allan Sandage in 1961.[10] In both these refinements, the fundamental criterion remained the shape of any given galaxy. Hubble's "tuning-fork" model uses the progressive flattening from ellipticals to spirals and highlights "form determined by rotational symmetry about dominating nuclei." An especially important add-on was the introduction of a transition stage: the lenticulars, a "bin" fit between ellipticals and spirals. Later schemes, and that of de Vaucouleurs in particular, included a better arrangement for the irregulars or Magellanic systems, as originally proposed by Lundmark. It is no surprise that Hubble's scheme has been at the root of the majority of the galaxy atlases that were subsequently published.

There were some dissenters with regard to the simplified view put forward by Hubble. "Too simple," many said. As we will see in this chapter, Halton Arp with his *Atlas of Peculiar Galaxies*, and to a lesser degree José Luis Sérsic with his *Atlas de galaxias australes*, deviated from Hubble's elegant scheme. Arp and Sérsic strongly emphasized "abnormal" morphologies, most often caused by an interaction or merger. These contrarians showed that galaxy atlases, as reference visual displays, were not only a way of identifying and selecting archetypes, but also a tool for highlighting complexity or deviants.

The following sections detail the main galaxy atlases published over the last 60 years (Fig. 10.2). The goal(s) or purpose(s) of each one are summarized, including the decision that went into their design, and their contents are briefly described. I show how these atlases of galaxies have been and continue to be "working objects" as defined in the Introduction. Chapter 11 deals more specifically with the users and the impact of these atlases. The reader just wishing for a quick overview of these atlases may concentrate on the first three and last one below, and then go straight to the next chapter.

A Compendium of Galaxy Atlases

1961. The Hubble Atlas of Galaxies, *Portraying Perfect Galaxies*

The American astronomer Allan Sandage (1926–2010) put together the lavishly illustrated mother of all atlases, *The Hubble Atlas of Galaxies*. Sandage had been an assistant of Hubble and he became one of the most influential astronomers of the twentieth century (Fig. 10.3). Sandage obtained his Ph.D. at Caltech in 1953, the year Hubble passed away. His supervisor was Walter Baade (Chapter 6). Published in 1961, this first great atlas is the embodiment and pillar of all galaxy atlases. Several generations of researchers and students have used the atlas extensively since then. The American astronomer Ronald Buta, lead author of *The de Vaucouleurs Atlas of Galaxies*, describes the impact of Sandage's atlas on his own professional trajectory: "The availability of this atlas at a price a high school student could afford brought galaxy morphology to an audience well beyond professional astronomers. Plus the quality of the reproductions and the general layout are so well done. The images are fascinating, and there was so much mystique connected with the different galaxy types

[9] G. de Vaucouleurs, Classification and Morphology of External Galaxies, *Handbuch der Physik*, 1959, Vol. 53, p. 275.
[10] A. R. Sandage, *The Hubble Atlas of Galaxies*, Washington: Carnegie Institution of Washington Publications, 1961, No. 618.

Fig. 10.2 Atlases of galaxies, a montage by Zoltan Levay and the author. With permission of Carnegie Science, University of Tokyo Press and Universidad de Córdoba, Observatorio Astronómico.

Fig. 10.3 Allan Sandage, 1973. Credit: Courtesy of the Archives, California Institute of Technology.

at the time. What the atlas showed captivated my imagination and directly contributed to my interest in ringed galaxies, the subject of my PhD thesis."[11] There is more on the impact of the atlas in the next chapter.

The atlas title carries the name of astronomer Hubble for a good reason. Edwin Hubble himself had initiated the atlas project to systematize his morphological classification of galaxies. As described in the previous chapter, Hubble presented his classification scheme at the International Astronomical Union Meeting in 1925. Soon after, he published it as a statistical investigation of 400 extragalactic nebulae in 1926.[12] In this highly cited work, Hubble used the inferred three-way morphology of galaxies to set the tuning-fork system, where normal and barred spirals merge with each other into the elliptical class (see Plate 9.1). Hubble's 1926 article influenced Harlow Shapley and Adelaide Ames when they put together their 1932 catalogue of the brightest nearby galaxies (Chapter 9).

At the dedication of the McDonald Observatory (Texas) in 1936, Hubble declared: "The structural forms of nebulae can be investigated with confidence because the necessary data are already available in the collection of direct photographs made with large reflectors." With Hubble's untimely death in September 1953, the classification project was left unfinished. As Hubble's intellectual heir, Allan Sandage inherited the unique collection of photographic plates of galaxies obtained by his mentor and other Mount Wilson observers. The 27-year-old Sandage did not delay and took up the reins of the extraordinary legacy. Using Hubble's notes, he re-inspected all the photographic plates that Hubble had used. Sandage re-asserted the main classification groups of Sa, Sb, Sc and Sd. He also refined the classification of the intermediary group (lenticulars) between ellipticals and spirals by using commentaries about E, S0, SB0 and Sa galaxy types that he gathered throughout Hubble's various notes and papers.

The photographic plates selected for the atlas came from the "magnificent set of plates" of galaxies brighter than photographic magnitude 13.0, north of declination minus 30°. These plates had been obtained with the Mount Wilson Observatory 60-inch and 100-inch telescopes between 1919 and 1948. To complete the new atlas, Sandage inserted the material from photographic plates just obtained with the new Mount Palomar 200-inch. A total of 176 separate galaxies were illustrated, accompanied by brief descriptions of observational data and numerous side comments. Sandage added a "variety" dimension to the Hubble system, and even included some "freak" cases.

As a working object, *The Hubble Atlas of Galaxies*, complete with its images, demonstrated the soundness and usefulness of morphology for classifying and understanding galaxies. It provided the visual forms or exemplars to apply the concept to other galaxies. The simple tuning-fork sketch was convivial and could be used effectively by a broad community of students and researchers.

Sandage had a grand exploration scheme in mind. He designed the atlas to explore the structural forms of galaxies as the initial step of a broad program of galaxy investigation.

---

[11] E-mail notes to the author, October 2013.
[12] E. P. Hubble, Extra-galactic Nebulae, *The Astrophysical Journal*, 1926, Vol. 64, pp. 321–369.

"*The Hubble Atlas* is but one stage in a developing program of nebular research," wrote Sandage in the introduction to the atlas. Indeed over his career, Sandage would complete and extend Hubble's work in a grandiose way, producing three additional atlases of galaxies published later. The best way to describe Sandage's vision is to quote the revealing sub-title, *Breaking the Code of Cosmic Evolution*, he gave to his superb book on the history of Mount Wilson Observatory.[13]

As I will demonstrate in the next chapter in describing the scholarly impact of galaxy atlases, *The Hubble Atlas of Galaxies* was an outstanding publication: it assembled the finest pictures of what had been seen as the world of galaxies and provided a guide on how to look at them in the proper way. Decades later, the value of Sandage's atlas is solidly recognized. MIT astrophysicist Alar Toomre emphasized "the extensive effort that Sandage made in that volume to construct at least a modest sense of order within this zoo."[14] But a "zoo" it remained for other researchers.

1966. Atlas of Peculiar Galaxies, *Picturing Roguish Galaxies*

The Hubble world of galaxies was too tranquil. The Swiss astronomer Fritz Zwicky (1898–1974), Russian astronomer Boris Vorontsov-Velyaminov (1904–1994) and, almost contemporaneously, Argentine astronomer José Luis Sérsic (1933–1943) had already pointed out strong morphological deviations from the elegant scheme initially proposed by Hubble.[15] It was no surprise that at least one leading astronomer was taken aback by Hubble's polished classification system. Every galaxy is "peculiar," vigorously wrote the American astronomer Halton Christian ("Chip") Arp (1927–2013) (Fig. 10.4). Along with Sandage, Arp had been a student of Walter Baade, at Caltech. He was also working at the Mount Wilson and Palomar Observatories, the same institution as Hubble and Sandage. Arp wanted to do something drastic to convey a different and more complete perspective of the world of galaxies: he produced another atlas, a very different one.[16,17] When published in 1966, Arp's *Atlas of Peculiar Galaxies* was received as a bombshell and even a frontal attack on the "authoritative" work of Hubble and Sandage. Arp trumpeted that the world of galaxies could and should be looked at in a very different way.

Several researchers had been uneasy with Sandage's apparent ignorance of galaxies showing signs of interaction. Even the "wrapped cigar" NGC 2685 or the obviously disturbed Messier 81, Messier M82 and NGC 520, all shown together on the same page in *The Hubble Atlas of Galaxies*, were considered "normal." Sandage's comment on NGC 520 is revealing as he wrote: "Inspection of the original plate suggests that NGC 520 is not a

[13] A. R. Sandage, *Centennial History of the Carnegie Institution of Washington, Volume 1: The Mount Wilson Observatory*, Cambridge: Cambridge University Press, 2004.

[14] A. Toomre, private e-mail to the author (16 October, 2014).

[15] See F. Zwicky, Multiple Galaxies, *Handbuch der Physik*, 1959, Vol. 53, p. 373.

[16] H. C. Arp, Atlas of Peculiar Galaxies, *Astrophysical Journal Supplement Series*, 1966, Vol. 14, pp. 1–20, and 57 plates of photographic reproductions. As noted in the abstract, "The Atlas was also available in large size, 11 x 14-inch photographic reproduction, from the California Institute of Technology Bookstore for a price of about $60 bound."

[17] A very nice reproduction of the original atlas as well as an extensive commentary is found in J. Kanipe and D. Webb, *The Arp Atlas of Peculiar Galaxies, A Chronicle and Observer's Guide*, Richmond: Willmann-Bell, Inc., 2006.

Fig. 10.4 Halton Arp at the Cassegrain focus of the Mount Palomar 200-inch telescope, preparing for observations with an image intensifier, 1967. Credit: Image courtesy of the Observatories of the Carnegie Institution for Science Collection at the Huntington Library, San Marino, California.

collision of two galaxies but rather a system of the M82 type."[18] Shrewdly, Arp characterized Hubble's scheme as an "idealization." Arp's contrarian atlas immediately occupied a

---

[18] A. R. Sandage, *The Hubble Atlas of Galaxies*, Washington: Carnegie Institution of Washington Publications, No. 618, 1961, pp. 41–42.

special place on astronomical library bookshelves, because it highlighted very vividly and with powerful images an obvious fact: that many galaxies in the present-day universe do not fit well under any of the Hubble-based classification systems (Figure 6.7 and Plate 10.1).

Where Was Arp Coming From?

Arp's perspective of the chaotic world of galaxies had already transpired in his review of the proceedings of the 13th Solvay conference on physics at the University of Brussels in 1964 that dealt with *The Structure and Evolution of Galaxies*. Arp felt that phenomena displayed by peculiar galaxies might be indicative of unknown physics. He was particularly impressed by the ideas of Armenian astronomer Victor Ambartsumian (1908–1996). Arp wrote then: "…Ambartsumian's introductory report, on the nuclei of galaxies and their activity, is outstanding. It courageously deals with the most important subject of all – the assumptions which underlie the description and attempted explanations of the phenomena. It is my opinion that what appear to be brilliant intuitions about ejection in galaxies, the role of nuclei, associations of spiral arms, blue objects, and quasars are really the result of Ambartsumian's considering the problem in great generality and, above all, of his careful reasoning coupled with the visual inspection and study of a very large number of actual galaxy forms."[19]

Arp was not the first to raise the issue of diversity and to emphasize chaos in the world of galaxy forms. For his atlas, Arp had selected objects from Vorontsov-Velyaminov's 1959 catalogue of 355 peculiar and interacting galaxies. The Russian astronomer and his colleagues had painstakingly identified weirdly shaped galaxies from the photographic National Geographic Society Palomar Sky Survey (PSS), which was completed with the Mount Palomar 48-inch Schmidt Telescope in 1956.[20] Re-inspecting the plates of the PSS used by Vorontsov-Velyaminov for his catalogue and the notes of A. G. Wilson, who obtained the original plates, Arp put together a new sample of galaxies. He then spent four years photographing the objects with the more powerful Palomar 200-inch; he also added to his sample objects from other observers. Of the 338 photographs shown in the atlas, most are from plates taken with the 200-inch telescope. Arp was making a point of using the same telescopes as Sandage and applying the best photographic processing techniques of the day.[21]

Arp's purpose was to produce his own atlas and present galaxies that had been ignored by Hubble and Sandage. It was almost an "anti-atlas." The work showed multiple examples of the various kinds of peculiar galaxies: perturbed, interacting, deformed and many others with various appendages or tidal tails (Fig. 10.5). The objects were grouped – based on a "rough, initial classification." It was an empirical approach, and the suspicious Arp wrote "the physical processes pictured are not understood." What Arp had in mind, then and

---

[19] H. C. Arp, Book Review: The Structure and Evolution of Galaxies, *Science*, 1966, Vol. 154, pp. 1439.

[20] B. A. Vorontsov-Velyaminov, R. I. Noskova and V. P. Arkhipova, *Atlas and Catalogue of Interacting Galaxies*, Sternberg Astronomical Institute, Moscow State University, Moscow, 1959. A second part was published in the 1970s.

[21] H. C. Arp and B. F. Madore later published *A Catalogue of Southern Peculiar Galaxies and Associations*, Vols. 1 & 2, Cambridge: Cambridge University Press, 1987.

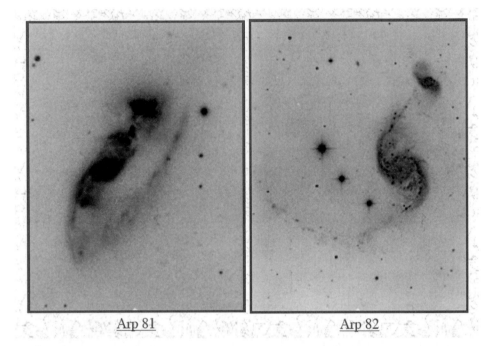

Fig. 10.5 Peculiar galaxies Arp 81 and Arp 82 from a page of Arp (1966), *Atlas of Peculiar Galaxies*. Courtesy of Carnegie Science.

thereafter, was that weirdo galaxies demonstrated the need for some novel kinds of physics. However, as the brilliant work of the brothers Alar and Juri Toomre showed in the 1970s, the unusual or peculiar galaxies had simpler explanations as interacting or merging systems.

Out of enthusiasm or frustration, Arp may have gone too far in seeing abnormality. As Sidney van den Bergh remarked, "some of the objects in Arp's catalogue are dwarfs that are not peculiar at all."[22] The large spiral Messier 101 (Arp 26) is asymmetrical at large galactocentric distances, but one may indeed wonder on how peculiar it is.

Notwithstanding it being a bit of a mixed bag, Arp's atlas became a fundamental contribution to galaxy structure. It highlighted a much broader range of morphologies than *The Hubble Atlas of Galaxies*. "Arp's atlas turned out to be an invaluable source of material during the subsequent development of the subject of galactic interactions and mergers."[23] The collection of peculiar galaxies had a tremendous impact as it presented galaxies as real dynamic systems, with the environment playing a major role in their evolution. As the Canadian astronomer Pierre Martin commented, "galaxies are just complicated beasts and not simple objects to be put in the box."[24]

[22] S. van den Bergh, *Galaxy Morphology and Classification*, Cambridge: Cambridge University Press, 1998.
[23] F. Schweizer, Observational Evidence for Interactions and Mergers, in *Galaxies: Interactions and Induced Star Formation*, R. C. Kennicutt Jr. et al. (editors), Saas-Fee Advanced Course 26, Swiss Society for Astrophysics and Astronomy, 1996, pp. 105–274.
[24] P. Martin, personal communication to the author, 2015.

Daston and Galison wrote: "the atlas aims to make nature safe for science; to replace raw experience – the accidental, contingent experience of specific individual objects – with digested experience."[25] This was Hubble's vision. Arp's vision was iconoclastic and he wished to destabilize established science. Today, the puzzle of galaxy shapes still shows gaps and cracks. Alar Toomre, one of the pioneers of computer simulations of galaxies and of their interaction, wrote recently: "Even today I cannot pretend that our 1972 work or its numerous and much more realistic sequels by others have reasonably explained any more than perhaps one-half of the strange images in Arp's atlas but, sad to say, we have collectively poured tons of cold water onto Chip's own hopes that perhaps/maybe/just-possibly he was on to some really new physics."[26]

As a working object, the *Atlas of Peculiar Galaxies* and its images illustrated a rival cosmological viewpoint. The morphology of galaxies was meaningful but could betray processes more fundamental and perhaps hide unknown physics. Arp's provocative approach invited the researcher and the student to keep a sharp eye out for unusual features as they might provide better insights into galaxy formation and evolution than the regular symmetrical normal galaxies. Peculiar objects might provide better clues to the physical forces responsible for shaping the observed forms of galaxies. Presenting exquisite photographical material he had obtained with the most powerful telescope of the time, Arp also conveyed a sense that he had access to rare and inaccessible categories of objects.

To understand the diverging approaches of Sandage/Hubble and Arp to galaxy structure, it is again useful to refer to another aphorism of Francis Bacon. "The greatest, and, perhaps, radical distinction between different men's dispositions for philosophy and the sciences is this; that some are more vigorous and active in observing the differences of things, others in observing their resemblances . . . Each of them readily falls into excess, by catching either as nice distinctions or shadows of resemblance."[27] As I will show, Arp and Sérsic were focusing on the differences, while Hubble, Sandage, de Vaucouleurs and their followers concentrated on resemblances. Today, we understand galaxy evolution much better following the pioneering work of the Toomre brothers and their followers. Most "peculiar" cases may be considered as transitional or temporary states in the complex lives of galaxies. Historically, such an epistemic clash is not new. Here's just one other example: the French naturalist Georges-François Leclerc (1707–1788), Comte de Buffon, was an ardent opponent of Carl Linnaeus' structured binomial classification system. A bit like Arp on galaxies, Buffon thought Linnaeus' approach tried to force disparate groups of species into "artificial assemblages," and that nature was richer than our systems.[28]

1968. Atlas de Galaxias Australes, *Illustrating the Normal and Abnormal*

Ready support for Arp's contrarian view soon came, from an astronomer working in the southern hemisphere. Until the mid twentieth century, all large telescopes were located

[25] L. Daston and P. Galison, The Image of Objectivity, *Representations*, No. 40, Special Issue: Seeing Science, 1992, p. 85.
[26] A. Toomre, private e-mail to the author (October 2014).
[27] F. Bacon, Aphorism 45, *Novum Organum*, 1620. See Hanover Historical Texts Project on-line.
[28] J. Elphick, *Birds, The Art of Ornithology*, New York: Rizzoli, 2014, p. 47.

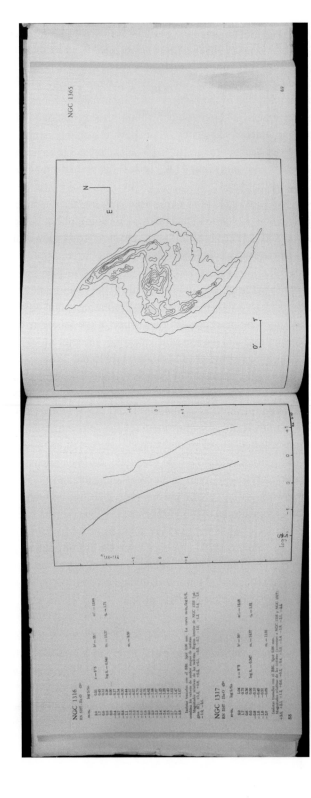

Fig. 10.6 Interior page of the *Atlas de galaxias australes*, showing the radial brightness profile (left) and the isophote map (right) of the barred galaxies NGC 1365. From Sérsic (1968), *Atlas de Galaxias Australes*. With permission of Universidad de Córdoba, Observatorio Astronómico.

mostly in the northern hemisphere. This disparity biased the observations as well as the sample of galaxies studied by researchers. The imbalance needed to be addressed, considering that the southern hemisphere hosts several archetypal objects, such as the Magellanic Clouds. The American astronomer Charles Perrine (1867–1951), whom we saw at Lick Observatory (Chapter 3), had moved to Argentina. While at Lick, Perrine had overseen the complete reconstruction of the Crossley 36-inch and had continued James Keeler's observational program on bright "nebulae." As director of the Observatorio Astronómico de Córdoba from 1918 to 1922, Perrine conducted a research program on galaxies and pioneered galaxy photography in the southern hemisphere using a small 76-cm telescope. Understanding what the instrument required for galaxy research, Perrine pushed for a larger reflector. On his initiative, a 1.54-m telescope was completed at Bosque Alegre, Córdoba, in 1942.

José Luis Sérsic followed brilliantly in Perrine's tracks, using the new and larger reflector. Initially, Sérsic intended to concentrate on studying the morphology and on the photometry of the galaxies brighter than magnitude 11.0 (Harvard scale) and south of declination −35° in the sky. He added more objects to assemble a sample large enough to produce a meaningful atlas. Sérsic's intention and ambition was to replicate for the southern hemisphere what Sandage had done so magnificently for galaxies observable from California.

Atlas de galaxias australes appeared in the Spanish language in 1968.[29] It was a fine work, rigorously produced. There were interesting similarities between Sérsic's atlas and Sandage's *The Hubble Atlas of Galaxies*: for example the format, the presentation and the use of the Hubble classification scheme were very similar. There were also some significant differences. For example, Sérsic insisted on obtaining and including in his atlas quantitative information, such as surface brightness (isophotal) maps and photometric profiles. John Reynolds had pioneered the measurement of a light-brightness profile across the bulges of spirals as early as 1913.[30] Sérsic did not have access to a very large telescope or a good site; hence, his photographs were not of the same quality, angular resolution and depth as the material of Sandage and Arp. Interestingly, Sérsic augmented the atlas with objects that revealed new phenomena or did not fit under the orderly scheme of Hubble and Sandage, hence the affinity with Arp's approach.

For Sérsic, morphology was important because of its "alto valor heurístico" (high heuristic value), which is a great aid for learning and discovering. His comments emphasize the nature of his atlas as a working object. He meant that images could reveal more than the schematic models, which were then "limited by our inability to resolve complex systems of equations." Nevertheless, he insisted that it is important to explore galaxies by employing more quantitative tools. In the footsteps of the 1920 seminal work of Reynolds,[31] Sérsic promoted and illustrated his use of photometry by plotting brightness profiles across galaxy images (Fig. 10.6), a prelude to the more systematic works by Japanese astronomers that

---

[29] J. L. Sérsic, *Atlas de galaxias australes*, Cordoba: Observatorio Astronómico, 1968.

[30] D. L. Block and K. Freeman, *Shrouds of the Night, Masks of the Milky Way and Our Awesome New View of Galaxies*, New York: Springer, 2008, p. 211.

[31] J. H. Reynolds, Photometric Measures of the Nuclei of some Typical Spiral Nebulae, *Monthly Notices of the Royal Astronomical Society*, 1920, Vol. 80, pp. 746–753.

were published later (see Fig. 10.8). To highlight the photometric material, Sérsic divided the atlas into two parts: (a) images taken with the 1.54-m of the Estación Astrofísica de Bosque Alegre, and (b) photometric maps, profiles and tabular data for several of the objects presented. Consequently Sérsic's work had a much more quantitative base than Sandage's atlas. This made the work unique and innovative for its time, and insured its durable impact.

Following the works of Arp (1966), Zwicky (1959) and Vorontsov-Velyaminov (1951), Sérsic proposed that galaxies that appear "abnormal" are an indication of some unusual and important processes. Thus, a distinctive aspect of the atlas was the evidence that galaxies undergo a much more complex and disturbed formation and evolution process than was apparently inferred by the Hubble–Sandage scheme. Illustrating this perspective, Sérsic's atlas included several examples of distorted and asymmetrical galaxies. Since anomalous galaxies are infrequent, these disturbances probably involved short timescales, e.g. 10^7 to 10^8 years. Commenting on Arp's term, "peculiar" galaxies, Sérsic asked: "What is a normal galaxy?" Perhaps wishing to be more constructive than Arp, Sérsic stated that "it is possible to clearly define the extreme of 'normality' and 'peculiarity,' leaving intermediary cases for discussion later." Sérsic aimed to be helpful and to advance the debate for the sake of better understanding galaxies and their evolution. As we will see in the next chapter, the surprising popularity of his atlas is testimony to the success of his broad vision and approach.

1981. The Revised Shapley–Ames Catalog of Bright Galaxies, *the Guide to All Galaxy Atlases*

In 1981 (a second edition appeared in 1987), Allan Sandage and Swiss astronomer Gustav A. Tammann published *The Revised Shapley-Ames Catalog of Bright Galaxies (RSA)*.[32] The *RSA* represented a complete overhaul of the original work by Shapley and Ames of 1932 and retained some of the format and the main content of their original work. However, this time, Sandage and Tammann added an important set of reference images of the main mor-phological types of galaxies, with new material on southern hemisphere objects obtained with the wide-field Irénée du Pont 2.5-m telescope of Las Campanas Observatory, Chile, that had been in operation since 1977. Harold Babcock played a leading role in having this telescope built, following a gift from Mr. and Mrs. Crawford H. Greenewalt to the Carnegie Institution of Washington. In their introduction to the atlas, Sandage and Tammann stressed the importance of galaxy surveys, first started by William Herschel, and they outlined the history of successive works. "The listing is meant as an aid in planning various observing programs based on knowledge of types as it existed in 1979." The addition of high quality images representative of the main galaxy types made the revised work a sort of mini atlas (Fig. 10.7).

An additional classification criterion was introduced. Sidney van den Bergh had shown that galaxies with the highest luminosities have the longest and most highly developed arms, whereas fainter systems show poorly developed arms. The luminosity classes range

---

[32] A. R. Sandage and G. A. Tammann, *The Revised Shapley–Ames Catalog of Bright Galaxies*, Washington: Carnegie Institution of Washington Publications, 1981.

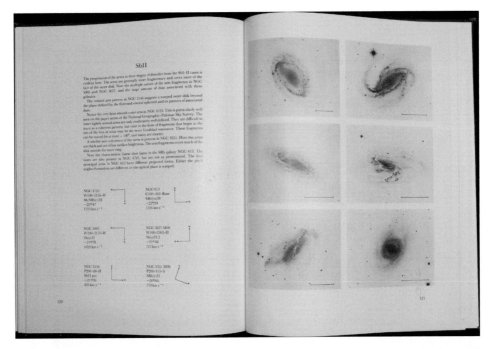

Fig. 10.7 Interior pages of *The Revised Shapley-Ames Catalog of Bright Galaxies*, Sandage and Tammann (1981). With permission of Carnegie Science.

from I to V, in decreasing luminosity; for example, a luminous spiral galaxy would have the extended descriptor Sbc I-II.[33] To make Hubble morphological types more inclusive descriptors, the *RSA* listing incorporated van den Bergh luminosity classes, called by Sandage the "beauty" criterion. American astronomers Debra Meloy Elmegreen and Bruce Gordon Elmegreen, who later developed a spiral-arm classification system of their own, highlighted this aspect, by introducing very visual descriptors such as "grand design" versus "flocculent" galaxies.[34]

As an important add-on of the *RSA*, the authors included a list of 822 new galaxies with photographic magnitudes of less than 13.2, which expanded on the original Shapley–Ames list. "Corrected Hubble types, complete coverage of the redshifts (adding to the earlier data by Humason and Mayall), and modern, mostly photoelectric magnitudes for every Shapley–Ames galaxy were listed."[35] In the later version of 1987, Sandage and Tammann provided new morphological types for about 200 southern galaxies and updated distances for many local galaxies. As a working object, the *RSA* advocated the Hubble–Sandage

---

[33] S. van den Bergh, A Preliminary Luminosity Classification of Late-Type Galaxies, and A Preliminary Luminosity Classification for Galaxies of Type Sb, *The Astrophysical Journal*, 1960, Vol. 131, pp. 215–223 and pp. 558–573.

[34] D. M. Elmegreen and B. G. Elmegreen, Arm Classifications for Spiral Galaxies, *The Astrophysical Journal*, 1987, Vol. 313, pp. 3–9.

[35] A. R. Sandage, *Centennial History of the Carnegie Institution of Washington, Volume 1: The Mount Wilson Observatory*, Cambridge: Cambridge University Press, 2004, p. 492.

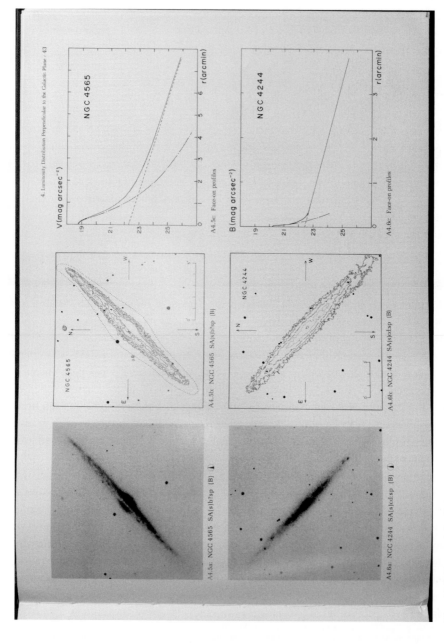

Fig. 10.8 Page of the Takase, Kodaira and Okamura photometric atlas showing its display of galaxy image, isophotal map and radial light profile. With permission of University of Tokyo Press.

classification scheme and provided the visual tools to apply it to a whole new set of galaxies. This catalogue and mini-atlas became an obligatory desktop reference and handbook for observational work, from planning to data reduction, analysis and interpretation. A limited number of archetypal objects and a concise format helped in memory recall, i.e. observers didn't need to have larger-sized atlases to hand. Nowadays the *RSA*, as a hard-copy book or e-book, continues to be a standard reference, just like a favorite recipe book.

1984. An Atlas of Selected Galaxies: With Illustrations of Photometric Analyses, *The Trust for Quantitative Imaging*

The first historical stage in atlas making, initiated by Hubble, was based on recognizing patterns from visual inspections of photographic images; it was essentially qualitative and subjective. Hubble's and Arp's pure morphological and qualitative approaches were considered insufficient by many. Wishing for a more quantitative approach, Japanese astronomers Bunshiro Takase, Keiichi Kodaira and Sadanori Okamura sought to follow in the footsteps of José Luis Sérsic.

Using the Japan-based Okayama 188-cm reflector and the Kiso Observatory 105-cm Schmidt telescope, Takase and his colleagues took photographs of galaxies through a set of colour filters, which were reduced to produce digitized photometric data with the help of high-performance measuring instruments and computerized image processing techniques. This photometric atlas in fact represented the first effort to include clusters of galaxies in a systematic photometric survey (Fig. 10.8).

They used the results of new digital photometry to classify galaxies. In their 1984 *An Atlas of Selected Galaxies: With Illustrations of Photometric Analyses*, the authors declared that galaxy classification required this new and more quantitative phase.[36] The atlas illustrated the computerized analysis of galaxy images using various image-processing techniques. The purpose was to demonstrate how photographs of galaxies could be reduced into quantitative data in the forms of calibrated isophotal maps, colour maps and radial luminosity profiles. The purpose of the small atlas was to demonstrate the technique of photographic-plate digitalization and computer analysis; it was not a systematic presentation of a large number of objects.

The intention of this quantitative analysis was to reduce the subjectivity and thus establish an "objective classification": the trio of Japanese astronomers claimed rightly that quantitative data were needed to provide reliable physical information on the luminosity and colour distributions for the different structural elements of galaxies. Although photometric data had been used in the earlier atlas of Sérsic, Takase et al.'s work marked the entry of galaxy atlases into the digital age, with a more advanced mathematical treatment that revealed hidden structural features of galaxy morphology. A highly innovative feature of the atlas was the application of two-dimensional Fourier analysis to quantitatively identify

---

[36] B. Takase, K. Kodaira and S. Okamura, *An Atlas of Selected Galaxies: With Illustrations of Photometric Analyses*, Tokyo: University of Tokyo Press and Utrecht: VNU Science Press, 1984.

and discriminate the modes of oscillations of galactic disks, which give rise to the various spiral patterns.

1988. Atlas of Galaxies Useful for Measuring the Cosmological Distance Scale, *Superlative Imagery*

The Hubble Space Telescope (HST) launched in 1990 was equipped (and still is) with cameras that provided a field of view of only a few arcminutes, a fraction of the angular size of the Moon, which is significantly smaller than that provided by most ground-based telescopes. Furthermore, this moderate-sized telescope operating in space held the extraordinary promise of an unprecedented angular resolution, which would reveal spatial details ten times finer than the best ground-based telescopes. This was a very significant jump in angular resolution, to the point that much more accurate visual references were required to see where the telescope was pointing and looking at. Among the several pre-launch preparations for the HST, an unusual venture was set up to help observers to find their way among the promised exquisite images: it was the production of a giant and finely illustrated atlas of galaxies. Designed and put together by Allan Sandage and John Bedke, the *Atlas of Galaxies Useful for Measuring the Cosmological Distance Scale* appeared in 1988.[37] The giant atlas, with 40 × 51-cm-sized pages, gets close to some of the largest-ever-published scientific atlases (Figs. 0.6 and 10.9).[38]

The completion of the Mount Wilson/Palomar/Las Campanas survey of galaxies by Sandage and collaborators in 1985 had provided most of the large-scale photographic material needed for this special atlas; some of it had already been used in the *RSA*. The sample of galaxies, to be potential HST targets, included 322 objects selected again from the Shapley–Ames set of bright galaxies. Sandage and Bedke stated that the purpose of this unusual atlas was to provide a selection of excellent galaxy photographs, with images of adequate scale for the resolution of the extragalactic stellar content with the powerful HST cameras.

Because the Virgo cluster of galaxies was to be a prime target of the HST programs, many objects from the central region of the Virgo cluster of galaxies were included in the new atlas. The last eight plates of the atlas showed several Virgo cluster galaxies, in anticipation of their use for determining the Hubble constant, i.e. the expansion rate of the universe, with greater accuracy; this was one of the reasons for building the HST in the first place. The ultimate theoretical goal was to obtain a more accurate determination of the spacetime geometry and the derivation of key cosmological parameters with higher precision. From these, the precise age of the universe could be inferred.[39]

To achieve this goal, a reliable and precise extragalactic distance scale, i.e. based on absolute distances, was required. The determination of distances to galaxies is based on the measurement of the properties of primary objects (luminous stars such as Cepheids, HII

---

[37] A. R. Sandage and J. Bedke, *Atlas of Galaxies Useful for Measuring the Cosmological Distance Scale*. Washington: Scientific and Technical Information Division, NASA, 1988.

[38] L. Daston and P. Galison (*Objectivity*, New York: Zone Books, 2007, pp. 23–25) describe James Bateman's double-elephant 17-kg and 68 × 98 cm folio atlas *The Orchidaceae of Mexico and Guatemala* (London: Ridgway, 1837–1843). This rare book can be found on sale currenly for close to $120,000 US dollars.

[39] Sandage and Bedke also use the expression "the Creation" or "since Creation" instead of the Big Bang.

Fig. 10.9 *Atlas of Galaxies Useful for Measuring the Cosmological Distance Scale*, by Sandage and Bedke (1988). Courtesy of NASA Scientific and Technical Information Division.

regions, supernova remnants) and comparing them with nearby Milky Way analogues for which distances are relatively well established. A high spatial resolution is required, and few galaxies are suitable. This is a result of a combination of the background light produced by the galaxy's billions of unresolved stars, the crowding together of the brighter stars, dust obscuration and a high inclination to the line of sight of the galaxy plane or just distance, all of which greatly reduce the ability to resolve individual stars. The new atlas would alleviate all these problems. It provided a set of galaxy photographs with images of an adequate scale for estimating how finely the stellar content could be resolved with the more powerful HST. The images of the atlas were assumed to be of good quality and were displayed at a scale suitable to help positioning HST observing fields for optimal control of stellar crowding. The technical objective was to resolve and recognize the largest possible number of stars, then to make use of the primary (Cepheid variables) and secondary targets (e.g. red and blue supergiant stars, novae, supernovae), the so-called distance indicators. The HST was expected to provide a spatial resolution gain of a factor of 10, and Sandage and Bedke illustrated this by putting images of the nearby galaxy Messier 33, and of Messier 101, located 10 times further away, side by side. As a working object, the atlas aimed to educate the potential users of HST ahead of time. The enormous size of the atlas still makes it a most conspicuous object in libraries and researchers struggle to find adequate space for it in their offices.

1988. The Color Atlas of Galaxies, *Searching for the Meaning of Colours in Galaxies*

Until the late 1980s, all galaxy atlases had used black-and-white images, reproduced from photographic plates. Today, we see so many spectacular colour images of galaxies that it is hard to imagine that, 30 years ago, producing colour photographs of galaxies was a real challenge. The American astronomer James D. Wray published the first *Color Atlas of Galaxies* in 1988.[40] It was an effort to produce true-colour pictures of galaxies with the highest fidelity. Wray used a meticulously chosen set of photographic emulsions and filters to ensure "true colours," as the author's "overriding consideration was the need for the colours to be reliable and interpretable." This meant reproducing a colour palette that would correspond to what a very sensitive human eye would perceive when looking at these galaxies.

The atlas comprised colour images of more than 600 galaxies of all forms, including "peculiar" and interacting ones. Wray assembled his core sample of galaxies from de Vaucouleurs et al.'s *Reference Catalog of Bright Galaxies* (RC1),[41] and added galaxies known or suspected to have nuclear activity or extended dust content. The imaging campaigns for the image database were conducted at McDonald Observatory in Texas, Las Campanas Observatory and Cerro Tololo Inter-American Observatory (CTIO) in Chile.

As the author surmised, "color images offer a source of aesthetic appreciation and scientific investigation." For example, colour allows one to distinguish a very young stellar population (bluish knotty regions) from old populations (extended yellowish light component).[42] To ensure colour authenticity, a colour reference was established, based on photometric data from individual bright stars in the field. The use of the Eastman three-colour dye transfer process enabled a correct colour balance at all brightness levels and a consistent relative surface brightness from galaxy to galaxy throughout the atlas. The manuscript submitted

---

[40] J. Wray, *Color Atlas of Galaxies*, Cambridge: Cambridge University Press, 1988.
[41] G. de Vaucouleurs, et al., *Third Reference Catalogue of Bright Galaxies*, New York: Springer, 1991.
[42] W. Baade had introduced the concept of stellar populations in his book *Evolution of Stars and Galaxies*, C. Payne-Gaposchkin (editor), Cambridge, MA: Harvard University Press, 1963.

by Wray was excellent. Despite all these efforts, however, the resulting product was disappointing because of the very poor quality of printing.

Nevertheless, the atlas benefited from the enthusiastic support of none other than Allan Sandage who emphasized that "the colors are true." Writing in the preface, Sandage envisaged that the work would become an exceptional pedagogical tool to teach three central themes: (1) the Hubble classification system, (2) Baade's stellar population concept, and (3) the life cycle of stars determining the stellar content of galaxies. In glowing words, Sandage continued: "From this book, galaxies will be chosen for work with the HST on the resolved stars for distance determination, applicable to the yet unsolved problem of the value of the Hubble constant; for work on the ground-based telescopes on star formation rates in galaxies of different Hubble types, studies of dust distributions, . . . " Sandage ended with unrefrained praise: "This atlas is important and . . . will be viewed some years hence as one of the major reference works produced in astronomy in the last half of the 20th century." Sandage had obviously reviewed the original manuscript with splendid photographic copies. Commenting on these originals, the American astronomer Harold G. Corwin Jr. writes, "The colors were not deeply saturated as most color astronomical photos seem to be, but were as subtle as I imagined the galaxies might be as viewed with a huge telescope."[43]

1990. Photometric Atlas of Northern Bright Galaxies, *Mapping Light across Galaxies*

In 1990, the Japanese astronomers Keiichi Kodaira, Sadanori Okamura and Shin-Ichi Ichikawa issued a new photometric galaxy atlas, the *Photometric Atlas of Northern Bright Galaxies*, based on photographic plates, obtained with the Kiso 105-cm Schmidt telescope, of 791 galaxies, chosen from a subset of the *RSA* listing.[44] Their work came in the wake of Takase et al. (1984), summarized above. They were continuing their colleagues' quantitative photometric approach. The new atlas went beyond just the demonstrative stage. It was part of a long-term project to produce a large database of the two-dimensional surface-brightness distributions of bright galaxies. It was a colossal enterprise, with images of all the galaxies recorded in the photographic V band (500–650 nm). The original photographs were all digitized using a Perkin-Elmer PDS photodensitometer, a complex and relatively expensive machine at the time, which converted the darkening level of the photographic emulsion into a quantitative measure of brightness.

Their effort was driven by the growing need to have increasingly accurate photometric data. Although not pursuing a complete survey, the authors' aim was to obtain homogeneous optical surface photometric data for as many nearby galaxies as possible, with isophotal maps and photometric parameters processed in a consistent way. From a scientific perspective, this atlas was to provide a quantitative galaxy database to be used as a reference for observations of galaxies conducted at other wavelengths, as new observing capabilities in the radio and X-ray domains began to provide maps and images of these objects (Chapter 7).

[43] H. G. Corwin Jr., e-mail communication with the author (December 2016).
[44] K. Kodaira, S. Okamura and S.-I. Ichikawa, *Photometric Atlas of Northern Bright Galaxies*, Tokyo: University of Tokyo Press, 1990.

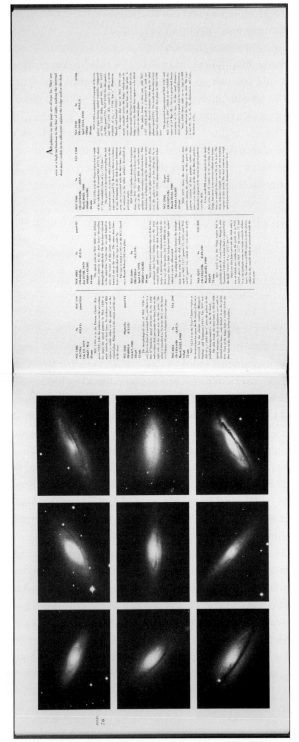

Fig. 10.10 *The Carnegie Atlas of Galaxies*, interior pages. With permission of Carnegie Science.

Kodaira, Okamura and Ichikawa produced a homogeneous and well-calibrated database of the largest quantitative sample of northern bright galaxies of its time. Each galaxy was presented in three panels: a photograph of each galaxy from the computer display screen; the isophotal map of the smoothed frame with about six isophotes at intervals of 1 magnitude per square arcsecond; and the surface-brightness profiles along the major and the minor axes. This was an impressive work that brought atlases of galaxies into the world of "big data." It was not until the Sloan Digital Sky Survey (SDSS), with its five-colour set of filters and greater sensitivity (due to the use of 2.5-m telescope and CCDs), that the scope and breadth of the Japanese atlas was surpassed.

1994. The Carnegie Atlas of Galaxies*, the Ultimate Book of Galaxies*

In 1994, Allan Sandage and John Bedke were back on the scene, and this time with a spectacular product. They published the epitome of all galaxy atlases, a grandiose two-volume publication. Initiated in the mid 1960s, *The Carnegie Atlas of Galaxies* represented over 25 years of effort.[45] Its original concept as the definitive scholarly atlas of galaxies had actually driven the construction of new telescopes, optimized for wide-field photography in the southern hemisphere. The project had also been enabled by the setting up of an important and unique photographic laboratory at the Space Telescope Science Institute in Baltimore, MD, the NASA Photolab. *The Carnegie Atlas of Galaxies* comprised images of 1,168 galaxies, compiling and illustrating most of the Shapley–Ames galaxies. The presentation was very much like the successful *The Hubble Atlas of Galaxies* of 1961.

According to Sandage and Bedke, the new atlas extended and refined *The Hubble Atlas of Galaxies* of 1961 and served as the definitive companion to the *RSA*. Its purposes were stated as being (1) a textbook of the *RSA* system of galaxy classification and (2) an illustrated compendium to aid in planning observing programs of bright galaxies, with both purposes fulfilling the criteria for an ideal scientific "working object."

With this ultimate atlas, the reader was to become familiar with the long-term program of galaxy imaging that Sandage and his colleagues had brought to completion. It had been a long adventure. The "Sandage project" had actually started in 1910 with the pioneering work of George Ritchey (Chapter 3). Over a period of 80 years, many photographic plates of galaxies had been obtained with the large optical telescopes in California and Chile. In the later decades of the twentieth century, new telescopes designed with wider-field capabilities (in particular the Chile-based Irénée du Pont 2.5-m telescope with its remarkable 2.2-square-degree field of view) helped Sandage and his collaborators to achieve the ambitious and stunning goal of imaging all the bright galaxies in the Shapley–Ames catalogue (Fig. 10.10). For Sandage, the atlas was the ultimate tool for "decoding cosmic evolution."

The introductory part of the atlas included some pedagogical musing on classification. "Can any classification be made independent of the classifier?" asked Sandage and Bedke. A quote from Francis Bacon (1620) is particularly interesting: "The human understanding,

---

[45] A. R. Sandage and J. Bedke, *The Carnegie Atlas of Galaxies*, Washington: Carnegie Institution of Washington Publications, 1994.

from its peculiar nature, easily supposes a greater degree of order and equality in things than it really finds." A warning that reminds us of Arthur Worthington's work and his setback on splashing liquid drops (Chapter 3).

For example, the idea of continuity or transition in the Hubble classification had been a fundamental question (and still is). "A purpose of this Atlas is to present evidence for the continuity. The ultimate purpose of the classification is to understand galaxy formation and evolution." Sandage and Bedke concluded, optimistically, "that the modern classification indeed describes a true order among the galaxies, an order not imposed by the classifier." The new atlas aimed to show that there was a stronger case for continuity along the sequence than could be made in 1961; continuity, but not evolution.

As will be seen in the next chapter, *The Carnegie Atlas of Galaxies* was extremely well received. It is an exemplary working object, meeting the goals initially set by Sandage in *The Hubble Atlas of Galaxies* 35 years earlier. The atlas is now available on-line, and it is widely used, due in great part to the large sample of galaxies of all classes. A later comment by Allan Sandage was almost an understatement: "With this atlas – an all-sky Carnegie project covering both celestial hemispheres – the vast project of photographing the galaxies, initiated by Pease and Ritchey in 1910, and continued by Hubble, Baade and Humason, into the 1950s, was brought to a close."[46]

2007. The de Vaucouleurs Atlas of Galaxies, *a Masterly Response*

Working somewhat independently of Sandage, French astronomer Gérard de Vaucouleurs had published a new galaxy classification system in the *Handbuch der Physik* in 1957 and 1959 (Chapter 9). The new system followed the broad categories of Hubble/Sandage but expanded them with finer details and subtle categories. Gérard de Vaucouleurs, assisted by his wife Antoinette and other colleagues, had classified several thousands of galaxies and published the results in a series of galaxy catalogues, e.g. *Third Reference Catalogue of Bright Galaxies (RC3)*. In this revised and expanded edition, de Vaucouleurs and collaborators presented 23,024 galaxies and about 18,000 had classifications. The reference catalogues were to some extent competitors to the *RSA*, but in several ways the works were complementary to each other.

However, de Vaucouleurs' classification scheme had never been well illustrated, certainly not to the level achieved for Hubble's scheme by Sandage in 1961, or the Hubble–Sandage revisions and updates of the mammoth atlases by Sandage and Bedke of 1988 and 1994. Although arriving late, Ronald Buta and his colleagues filled the gap most expertly. In 2007, the researchers came up with a wonderful galaxy atlas: *The de Vaucouleurs Atlas of Galaxies*, published in 2007, is the latest of the great atlases of galaxies.[47]

It is a masterpiece. The authors followed a hint from the astronomer Ivan King who suggested that the quality of the illustrations was the likely reason why the classification

---

[46] A. R. Sandage, *Centennial History of the Carnegie Institution of Washington, Volume 1: The Mount Wilson Observatory*, Cambridge: Cambridge University Press, 2004, pp. 490–491.

[47] R. J. Buta, H. G. Corwin Jr. and S. C. Odewahn, *The de Vaucouleurs Atlas of Galaxies*, Cambridge: Cambridge University Press, 2007.

scheme of the *RSA* has been so popular. So Ronald Buta, Harold Corwin Jr. and Stephen Odewahn recreated for the de Vaucouleurs classification scheme what Sandage et al. did for the Hubble classification scheme: they illustrated the types and the meanings of the class notation by using large-scale images obtained with large ground-based telescopes, all of which were carefully annotated.

Paraphrasing de Vaucouleurs' perspective on galaxy shapes, the authors stated that one should avoid seeing sharp boundaries between the morphological classes of the galaxies. This is the same idea of "continuity" emphasized by Sandage and Bedke. Buta et al. showed that the structures of galaxies are wide-ranging, with a continuum of forms, requiring a classification system not only of precision but also of great flexibility.[48] For example "the intermediate bar classification SAB is one of the hallmarks of the de Vaucouleurs system, and is used to recognize galaxies having characteristics intermediate between barred and non barred galaxies."[49]

In a noticeable technical difference and innovation, Buta and his collaborators used modern CCD electronic detectors to image the galaxies, instead of the traditional photographic plates. It was the first major and systematic classification of galaxies based on digital images. An obvious advantage of CCDs is the greater dynamic range, allowing features of highly different contrast to be seen and measured on the same image. They also completed their illustrative work with the benefit of a solid historical perspective and mature understanding of galaxies that had been acquired in the 50 years since de Vaucouleurs had proposed his morphological types.[50]

The authors also described carefully how the de Vaucouleurs scheme differed from the Hubble–Sandage revised types. Instead of using the *RSA* as a sample base, Buta et al. tapped the general database of images from the astronomical community. Of the 523 images, very few were intentionally taken for the atlas, in sharp contrast to all other atlases presented in this chapter. As a working object, the atlas demonstrated the reliability and advantages of digital images for galaxy classification and magnificently illustrated the de Vaucouleurs revised classification scheme. It also contains one of the most complete and finest introductions to galaxy classification. The work is accompanied by a galaxy morphology website, with the images of the atlas plus additional ones covering as wide a redshift as possible.

Citing examples from the space-based Spitzer Space Observatory, which had provided images at infrared wavelengths from 3.6 to 8.0 microns, and the satellite GALEX in the ultraviolet, Buta et al. wrote "there is no question at this time that galaxy morphology is a vibrant subject." The authors re-emphasized that the field of galaxy classification had finally emerged from its pre-1990s second-class status, when it was viewed with disdain, probably because it was too qualitative and somewhat arbitrary.

---

[48] See for example: J. Kormendy, Observations of Galaxy Structure and Dynamics, in *Morphology and Dynamics of Galaxies, Proceedings of the 12th Advanced Course*, Saas-Fee, Switzerland: Observatoire de Genève, 1983, pp. 113–288; and R. Buta, Galaxy Morphology, in *Planets, Stars, and Stellar Systems*, Volume 6, W. C. Keel (editor), Dordrecht: Springer, 2011.

[49] R. Buta, et al., Mid-infrared Galaxy Morphology from the Spitzer Survey of Stellar Structure in Galaxies (s4G): The imprint of the de Vaucouleurs Revised Hubble–Sandage Classification System at 3.6 micron, *Astrophysical Journal Supplement Series*, 2010, Vol. 190, pp. 147–165.

[50] G. de Vaucouleurs, Classification and Morphology of External Galaxies, *Handbuch der Physik*, 1959, Vol. 53, p. 275.

The new atlas was also didactic, using galaxy morphology to explain their formation and evolution. The Hubble sequence "did not always exist but was built up over time via mergers or secular evolution or both." When did it all fall into place? The authors indeed proposed a clear answer: between redshift $0.5 < z < 2$, i.e. when the universe was between one third and two thirds of its present age.[51]

This atlas brings to a conclusion my review of the main galaxy atlases. Nearly half a dozen more specialized galaxy atlases were published during the same period, which are discussed in the appendix.

New Trends in Galaxy Imaging

If we count "A Survey of the External Galaxies Brighter Than the Thirteenth Magnitude" published by Harlow Shapley and Adelaide Ames in 1932, as the start of galaxy atlases, there have been about 16 such atlases, including some more specialized atlases (see appendix), over a period of about 80 years, corresponding to a rate of about one atlas per five years. It is also remarkable that almost all these atlases used material from photographic, emulsion-covered glass plates.

Only two research atlases, Hickson's *Atlas of Compact Groups of Galaxies* (see appendix) and Buta et al.'s *The de Vaucouleurs Atlas of Galaxies*, used modern CCD electronic cameras to produce images of galaxies for their atlases. This may appear somewhat surprising since CCDs have been around for the last 25 years. In his review of *The Carnegie Atlas of Galaxies*, Brian Skiff commented on the "mature technology of photographs," and rightly warned that Sandage and Bedke finished their atlas "just in time before the demise of the photographic plate. . . . It will almost certainly be the last of major photographic atlases of large-scale astronomical images to be published." This statement appeared to imply that CCD cameras (then offering relatively small fields) would not be able to produce large images. It has taken some time for CCDs to be in formats large enough to image galaxies in an effective way. But this situation has dramatically changed, with the current manufacturing of large-sized CCD chips and techniques that allow their mosaic assembly in CCD cameras with a field of view now larger than even the largest photographic plates used by Sandage and others.[52]

Nevertheless, compilers of galaxy atlases of the previous decades had good reason to favor photographic plates as they were of sufficient size to capture the large fields of view of the telescopes. It would actually be very interesting to see a replication of Allan Sandage's *The Hubble Atlas of Galaxies* or of James Wray's *Color Atlas of Galaxies* with images obtained with the modern electronic imagers on the HST and the large, ground-based fine-imaging telescopes such as the Very Large Telescope (VLT), Gemini and Subaru.[53]

---

[51] Redshift (z) is a widely used unit of distance based on the expansion of the universe; z is measured by spectroscopy. It is defined as follows: $1 + z = \lambda_{obs}/\lambda_{lab}$, where λ_{obs} is the observed wavelength of a cosmic source, and λ_{lab} is the laboratory wavelength of the same spectral line. The larger z is, the further away is the source and the further in time we are looking.

[52] The Ritchey–Chretien 2.5-m Irénée du Pont Telescope at Las Campanas Observatory in Chile had a photographic camera with 50×50-cm-sized plates, which fully exploited the 2.1-degree field of view.

[53] The giant HyperSuprimeCam on the Subaru 8-m telescope feeds a 116-CCD array (60 cm in diameter) with 870 mega pixels covering a 1.5-degree field of view.

Atlases of galaxies have also undergone a more subtle evolution. Martin J. S. Rudwick has described the increasing importance of visual content vis-à-vis text in geological publications of the nineteenth century and showed this to be mainly the result of evolving printing technologies.[54] Looking at the successive atlases of galaxies, an interesting trend can be noted: the first generation of atlases (those published before 1990) are dominated by large-sized images accompanied by very concise, almost telegraphic, text; see, for example, Sandage's *The Hubble Atlas of Galaxies*, or Takase, Kodaira and Okamura's *An Atlas of Selected Galaxies*. In contrast, the images of *The Carnegie Atlas of Galaxies* (1996), while still filling the main content of each page, are accompanied by a much longer and descriptive text. The most recent *The de Vaucouleurs Atlas of Galaxies* (2007) has the highest ratio of space for text relative to image per page. It is almost as if, as more and more complexities and details of galaxy morphology are being revealed, authors feel they have to say more, because the images are less self-evident for the non-expert.

Although their classification has been driven mainly by observations in the visible part of the spectrum, those in the infrared have reinforced the traditional approach to sorting galaxies. Galaxy classification is a dynamic area of research and is brimming with new developments.[55] The availability of large-format infrared imagers also opens new possibilities. This has been well demonstrated by Ron Buta and his colleagues with their mid-IR classification of galaxies using the Spitzer Space Observatory Infrared Array Camera (IRAC) images of more than 200 galaxies at a wavelength of 3.6 microns. The Spitzer Space Telescope is a small telescope (with mirror of 85 cm in diameter) flown in an Earth-trailing orbit around the Sun. It was cooled to 5.5 K and had instruments that produced images and obtained spectra at wavelengths between 5 and 40 microns (see Fig. 0.4). Having used up its active cooling liquid helium, the telescope is still operational in the shorter wavelengths although with limited capabilities. The observatory was named after Princeton University astrophysicist Lyman Spitzer (1914–1997), who first proposed the idea of telescopes operating in space in 1946. Buta et al.'s key finding is "that 3.6 micron classifications are well correlated with blue-light classifications, to the point where the essential features of many galaxies look very similar in the two very different wavelength regimes. Drastic differences are found only for the most dusty galaxies." The advantage of infrared imaging is that it is less prone to the "hiding" effect of obscuring dust, an issue that Heber Curtis had already addressed in 1918 (Chapter 3).

And what about the Galaxy Zoo project? This is a citizen science project that was started in order to classify the millions of galaxies imaged by the Sloan Digital Sky Survey.[56] Classifying this multitude of objects by the naked eye appeared impossible. There is no computer algorithm that can reproduce the human ability to tell galaxy classes apart. Also, powerful new telescopes and techniques have enabled galaxies to be observed further back in time. Very deep images reveal a world of galaxies, which seems different from those in

---

[54] M. J. S. Rudwick, The Emergence of a Visual Language for Geological Science 1760–1840, *History of Science*, 1976, Vol. XIV, pp. 149–195.

[55] R. Buta et al. present a detailed discussion of recent developments in galaxy classification in the introductory part of *The de Vaucouleurs Atlas of Galaxies*, op. cit., pp. 27–80.

[56] To learn more on the Galaxy Zoo project and the Sloan Digital Sky Survey, see www.galaxyzoo.org and www.sdss.com.

the nearby universe, even taking into account the reduced spatial resolution on the objects. Despite present obstacles, it is clear that large data sets and artificial-intelligence tools will advance galaxy classification in ways we cannot predict.

It is important to reiterate that galaxy classification systems are empirical; they are not based on theory. Nevertheless, enormous effort has been devoted to linking the shapes of galaxies to the physical processes taking place through the aeons of galaxy history.[57] Allan Sandage has summarized some key points for understanding how galaxies appear to us: "The difference in time variation of the star formation rate for galaxies of different Hubble type seems capable of explaining the five principal facts of observation concerning the Hubble sequence. These are (1) the bulge-to-disk ratio is a function of Hubble type, (2) the disk surface brightness varies systematically along the sequence, (3) the integrated colour also changes progressively as a monotone function of the type, (4) the mean age of the disk also varies in the same progression, and (5) the present star formation rate per unit mass for Sc galaxies is much larger than for S0 and Sa galaxies."[58]

What is astonishing is how successful morphology has been as a classification criterion. In his book on the topic of discovery and classification, the American astronomer Steven Dick commented that it was remarkable that galaxy classification systems proved to be as good as they were, given the difficulties in correlating galaxy morphology with physical significance.[59] The simple property of galaxy shape is what runs through all the atlases I have presented and discussed. Now as we probe the distant universe and try to classify young galaxies, it is astonishing that we are using the simple criteria developed in the 1920s by Reynolds, Lundmark and Hubble. Halton Arp would be very pleased by the multitude of "peculiar" objects observed in the young universe . . . Galaxy atlases have been an outstanding manifestation of the power of images within scientific discovery. Let us see how they have been used.

---

[57] Recent development in galaxy classification includes theoretical aspects and insights. See R. Buta, Galaxy Morphology, in *Secular Evolution of Galaxies*, J. Falcón-Barroso and J. H. Knapen (editors), Cambridge: Cambridge University Press, 2013, pp. 155–258.

[58] A. R. Sandage, Star Formation Rates, Galaxy Morphology, and the Hubble Sequence, *Astronomy & Astrophysics*, 1986, Vol. 16, pp. 89–101 (quote from p. 89).

[59] S. J. Dick, *Discovery and Classification in Astronomy, Controversy and Consensus*, Cambridge: Cambridge University press, 2013, p. 260.

11

Atlases of Galaxies, Viewed by Their Users

> The atlases drill the eye of the beginner and refresh the eye of the old hand.
>
> *Lorraine Daston and Peter Galison*[1]

> An artistic achievement, this volume is also a scholarly work of major proportions.
>
> *George Field*[2]

> For amateurs, this is a marvelous photographic atlas that includes almost all the galaxies they usually look at.
>
> *Walter Scott Houston*[3]

Who Have Used Galaxy Atlases and What Have They Used Them For?

The atlases of galaxies published during the last 60 years or so have been scholarly creations. They have reflected our rapidly increasing knowledge and understanding of galaxies. These special publications were also driven by progress in our ability to image the sky with larger, more powerful telescopes and cameras of increasing sensitivity and larger fields of view. It is well worth quoting again Daston and Galison: "Not only do images make the atlas; atlas images make the science." Atlases are the repositories of recorded images for the observational sciences. Let us explore the following question: beyond being displays and repositories of current knowledge, do they also generate new knowledge?

Scientific atlases are considered to be specialized products, aimed at the experts. Hence, galaxy atlases were produced primarily to fulfil specific needs of researchers in the discipline, and to inform the practitioners in adjacent research fields. Nonetheless, a few of the galaxy atlases discussed in the previous chapter (and in the appendix) have reached a wider audience. Astronomy is not that exclusive a discipline. Like ornithology, entomology and other natural sciences, it sustains cohorts of amateur astronomers around the world. Many

---

[1] L. Daston and P. Galison, *Objectivity*, New York: Zone Books, 2007, p. 23.
[2] G. Field, Book Review: The Hubble Atlas of Galaxies, *American Scientist*, 1962, Vol. 50, p. 212A.
[3] S. H. Walter, Book Review: Atlas of Galaxies Useful for Measuring the Cosmological Distance Scale, *Sky and Telescope*, 1989, Vol. 78, p. 40.

of them are curious and inquisitive, and several are skillful astrophotographers. Over the years, passionate aficionados have equipped themselves with the latest technology to make their telescopes and imagers more powerful, to satisfy their quest for more knowledge. With their ability to observe and photograph thousands of galaxies, even with relatively small telescopes, amateur astronomers have been keen users of galaxy atlases.

This increasing audience for fine astronomical images and atlases has resulted in the recent publication of a few atlases aimed at a wider readership.[4] For the dedicated amateur, Jeff Kanipe and Dennis Webb, with a team of amateur astronomers, have put together the unique *The Arp Atlas of Peculiar Galaxies*.[5] For the general public, a fine example is Christensen et al.'s *Cosmic Collisions: The Hubble Atlas of Merging Galaxies*.[6] And Wray's *The Color Atlas of Galaxies* and, more recently, *The Multiwavelength Atlas of Galaxies* by Glen Mackie (see appendix) have a strong pedagogical approach.[7]

Several scientific atlases have been used for purposes beyond the original intent of their designers and authors. This has been in great part due to the high quality of the publications and to the beauty of the images presented in these works. Some of the galaxy atlases proved to be no exception to this broader usage. By carefully browsing through *The Hubble Atlas of Galaxies*, even a newcomer can quickly learn about the major classes of galaxies and recognize the basic morphologies, after a brief initiation. Ultimately, the majority of the galaxy atlases are pleasing to the eye and rewarding for the inquisitive mind; the fact that the images are black-and-white reproductions has not diminished the aesthetic value of the books. Finally, in the fine tradition of atlases, size counts. The designers and producers of galaxy atlases took care to include several full-page reproductions of good quality, some awe-inspiring.

Consequently, atlases have served for training and teaching, and to initiate the neophyte researcher. In teaching, atlases have been used to define the standards of how galaxies are to be seen and depicted. Teachers have used the objects illustrated to select a galaxy or a set of galaxies they wish to present to students of an astronomy course; or to highlight a distinct category of objects or a single object that deviates from the "standards" of the atlases; or even point out objects (e.g. "peculiar" galaxies) that are blatantly omitted from the atlas. Canadian astronomer Marshall McCall writes: "I have used *The Hubble Atlas of Galaxies* in pretty much every course I teach, because it is a foundation for classifications. However, I use it to illustrate the Revised Hubble System of de Vaucouleurs, not the Hubble System of Sandage. Another book I use for teaching is this giant *Atlas of Galaxies* produced by NASA."[8] On many occasions, atlases have been the source of material to define a sample of objects for a thesis project and to initiate a new research program.

---

[4] Not an atlas, but a highly popular large book of galaxy images, is T. Ferris's *Galaxies*, New York: Random House Publishing, 1988.

[5] J. Kanipe and D. Webb, *The Arp Atlas of Peculiar Galaxies, A Chronicle and Observer's Guide*, Richmond: Willman Bell, Inc., 2006.

[6] L. L. Christensen, D. de Martin and R. Y. Shida, *Cosmic Collisions: The Hubble Atlas of Merging Galaxies*, New York: Springer, 2009.

[7] J. Wray, *Color Atlas of Galaxies*, Cambridge: Cambridge University Press, 1988; G. Mackie, *The Multiwavelength Atlas of Galaxies*, Cambridge: Cambridge University Press, 2011.

[8] E-mail exchange with the author.

Citations of Atlases and Their Usage

An accepted measure of the use, success and importance of a scientific work is the cumulative number of citations it has received over the years, especially in the specialized articles published. There may be various reasons for citing a galaxy atlas: authors may simply wish to refer to an image in the atlas or give a description of a chosen sample of galaxies for observing or study; they may be comparing an object with the archetypes in the atlases; or they may also be discussing classification schemes, with respect to their success, limitation or failure.

Table 11.1 lists the galaxy atlases covered in this book, with the year in which they appeared. The successive columns indicate whether photometric information (isophotal maps, radial brightness profiles) is included; the type of audience that the atlas was mainly aimed to reach, or has reached (research, training or public); the number of citations (from SAO/NASA Astrophysics Data System) up to the beginning of 2017; and the number of book reviews is given in the last column.

In the fields of astronomy and astrophysics, and the physical sciences in general, books (including atlases) are not cited as systematically as research articles in peer-reviewed journals. Their relative importance for citation is less than that in the humanities or social sciences. This is not because books are of lesser importance but reflects a cultural habit within the natural sciences' research communities. Here, books are thought to be compendia or summaries of the research published in the scholarly journals, hence used in teaching or researching fundamentals, e.g. in exploring a new avenue of research or for the better understanding of a problem being addressed. The original articles in the professional journals are cited in preference because they contain more details, have the fully listed authorship and institutional affiliations, and enjoy the prestige of the journal that published them. Therefore, in terms of citations, atlases may be regarded more as books than original research articles.

Since the majority of the galaxy atlases were published some time ago, the number of citations is still a meaningful measure of their impact, especially for those that were published before 2000. It is noticeable that Sérsic's *Atlas de galaxias australes*, published in the Spanish language and not a particularly flamboyant publication, had about 1,340 citations at the beginning of 2017; it is, together with Sandage's *The Hubble Atlas of Galaxies* (1,263 citations) the most cited atlas of galaxies, followed by Arp's *Atlas of Peculiar Galaxies* (1,141 citations). The popularity of Sérsic's work is very likely due to its inclusion of photometric maps and data. However, the *RSA* mini-atlas/catalogue had the greatest number of citations, 1,587.

Moreover, as it is well known that atlases of galaxies reach beyond the specialists, it is a reasonable to assume that they have been used several times this number of citations. Whilst being a quantitative measure the actual number of citations does not necessarily give the full intellectual impact of a published work. For example, it is impossible to measure the educational input on a young kid who happens to open the pages of an atlas and "gets hooked" by the images while visiting a friend. And how many Ph.D. students were

Table 11.1 *Atlases of galaxies, book reviews and citations*

Atlas title	Authors	Year	Isophotes	B&w/colour	Research	Training	Public	Citations	Book reviews
"A Survey of the External Galaxies Brighter Than the Thirteenth Magnitude"	Shapley and Ames	1932	No	Catalog only	✔			64	0
The Hubble Atlas of Galaxies	Sandage	1961	No	B&W	✔	✔	✔	1,263	4
Atlas of Peculiar Galaxies	Arp	1966	No	B&W	✔	✔		1,141	0
Atlas de galaxias australes	Sérsic	1968	Yes	B&W	✔	✔		1,340	1
An Atlas of 21 cm HI Line Profiles of 61 Galaxies	Rots	1979	Yes	Profiles, maps	✔			5	
The Revised Shapley–Ames Catalog of Bright Galaxies	Sandage and Tammann	1981, 1987	No	B&W	✔	✔		1,587	3
An Atlas of Selected Galaxies	Takase, Kodaira and Okamura	1984	Yes	B&W	✔			14	
Color Atlas of Galaxies	Wray	1988	No	Colour	✔	✔	✔	67	4
Atlas of Galaxies Useful for Measuring the Cosmological Distance Scale	Sandage and Bedke	1988	No	B&W	✔	✔		139	4
Photometric Atlas of Northern Bright Galaxies	Kodaira, Okamura and Ichikawa	1990	Yes	B&W	✔			42	
Atlas of Compact Groups of Galaxies	Hickson	1993	?	B&W – CCD imaging	✔	✔		109	7
The Carnegie Atlas of Galaxies	Sandage and Bedke	1994	No	B&W	✔	✔	✔	310	4
The de Vaucouleurs Atlas of Galaxies	Buta, Corwin and Odewahn	2007	No	B&W – CCD imaging	✔	✔		60	3

re-oriented in their research by flipping through the pages of one of these great galaxy atlases? This is an impact we learn only by interviewing active researchers.

Inspiring Atlases

As has been mentioned several times already, the impact of Sandage's *The Hubble Atlas of Galaxies* has been huge; it has inspired many young researchers. Together with the *Atlas de galaxias australes* and *The Carnegie Atlas of Galaxies*, it has served legions of students and researchers trying to understand and explain morphology or to account for the "tuning-fork" classification: How does one interpret the observed silhouette of a galaxy? How do galaxies form and evolve? What gives shape to galaxies? This questioning is epistemologically fruitful because several observed properties of galaxies correlate with shape. Shapes of galaxies matter as they tell us about the origin, formation process and dynamic states of galaxies. Thus, galaxy atlases, in presenting collections of shapes, are revealing compendia of key information.

The Hubble Atlas of Galaxies and others that followed became engines for defining new research programs. Sandage himself kept expanding his survey of galaxies beyond the original project of Hubble. For example, by exploiting the wide field of view of the Irénée du Pont telescope in Chile, he conducted a survey of the Virgo cluster, obtaining 67 large 50×50-cm plates to cover an area of about 140 square degrees of the sky.[9] The survey yielded a catalogue of 2,096 galaxies, most of which are Virgo cluster members. It was used in particular to refine the classification of dwarf galaxies.[10] Below are given a few more examples of such research programs.

The work of the American astronomer Sandra Faber on elliptical galaxies is a fine example. Because of their differing colours and gas content, one may think that spirals are relatively young and ellipticals, old. In disentangling their mass-to-light ratios, stellar populations, age and metallicity, Faber demonstrated that stellar populations in many ellipticals are surprisingly young.[11] Or the work of the French astronomer Françoise Combes together with the Swiss theoreticians Daniel Pfenniger and Louis Martinet on the secular evolution of spiral galaxies, where disk galaxies may go through successive episodes of bar formation followed by destruction, the central bulge of the galaxies growing each time through scattering the orbits of stars. Disk galaxies would swing from one branch of the tuning fork to the other: normal to barred and barred to normal, growing a bigger bulge. On a grander scale, some spirals would evolve dynamically from Sm to Sa through bar formation and destruction, bulge growth and mergers.[12] These works aimed at explaining the Hubble sequence, as illustrated in atlases of galaxies.

---

[9] The Irénée du Pont telescope was the result of a gift in 1970 from Mr. and Mrs. Crawford H. Greenewalt to the Carnegie Institution of Washington. Mrs. Crawford H. Greenewalt was the daughter of Irénée du Pont, the founder of the Dupont Company.

[10] Allan Sandage published the work on the Virgo cluster in six papers between 1984 and 1987 as collaborative works with Gustav Tammann and Bruno Binggeli.

[11] The *Faber–Jackson relation* between the luminosity and the central stellar velocity dispersion of elliptical galaxies is a fundamental empirical relation. It can be used for determining distances to external galaxies.

[12] D. Pfenniger, L. Martinet and F. Combes, Secular Evolution of Galaxy Morphologies, 1996, arXiv:astro-ph/9602139v2

For the Australian astronomer and cosmologist Kenneth Freeman, *The Hubble Atlas of Galaxies* was the main atlas available early in his career and it certainly influenced his work. With the images covering most types of galaxies and the texts full of Sandage's wisdom on the subject, much of Freeman's views about the nature of disk galaxies came from looking at the atlas images of edge-on disks. Freeman was later intrigued by Arp's *Atlas of Peculiar Galaxies*, and spent much time looking at its images. Piqued by the idea of galaxies inter-acting with the intergalactic medium, Freeman saw a set of galaxies of the atlas (Arp 118, 142, 144, 147) as if they were being stripped while ploughing through the intracluster hot gas (Plate 7.7). Sérsic's *Atlas de galaxias australes* served a different function. "The pictures are not so great, but the atlas includes photographic surface photometry (and I think it may have introduced the Sérsic law). Maybe that is why it is so well cited. I used the Sérsic atlas a lot when I was working on the properties of the exponential disk, because there was not so much quantitative surface photometry available at the time."[13] The Sérsic profile describes mathematically how the intensity of light of a galaxy varies as a function of distance from its center.

For the Canadian astronomer Wendy Freedman, a principal investigator of a large project on the Hubble Space Telescope (HST), Sandage and Bedke's *Atlas of Galaxies Useful for Measuring the Cosmological Distance Scale* was "enormously helpful in preparing for the Key Project and many of us spent many, many hours pouring over those images to choose our target fields."[14] For quite a few other researchers, the most immediate use of the galaxy atlas images has been to compare against other images of the same galaxy to check for suspected supernovae, by making sure that the suspected supernova is not a nearby star hiding in the galaxy background light.

The select classification of barred galaxies by de Vaucouleurs was the inspiration, for Pierre Martin, to distinguish and characterize bars using quantitative parameters, such as bar length, ellipticity and torque. *The Carnegie Atlas of Galaxies*, on the other hand, was an essential tool during his postdoctoral years. The superb images helped in identifying objects of choice to observe at the telescope, in order to reveal the impact of bars in galaxy evolution.

Arp's *Atlas of Peculiar Galaxies* had an enormous influence on the learning and under-standing of the structure and evolution of galaxies. It inspired the early work of the Estonian–American astronomers Alar and Jüri Toomre, who were among the very first researchers to carry out successful computer simulations of galaxy interactions and tidal tails. Arp's work and the Toomres' simulations gave fuel to the debate about the respective role of nature and nurture in galaxy evolution.[15] Arp and the Toomres showed us that our telescopes were catching galaxies in the act of transformation.

Indeed, the Toomres' computer simulations opened up a tremendously productive approach; they invited galaxy researchers to look at familiar but oddly shaped galaxies with new eyes. It became obvious that galaxies did not always evolve passively nor isolated from

---

[13] Private e-mail note to the author (November 2014).   [14] Private e-mail note to the author (October 2014).
[15] A. Toomre and J. Toomre, Galactic Bridges and Tails, *The Astrophysical Journal*, 1972, Vol. 178, pp. 623–666.

each other. There were deviants from Hubble's ideal shapes shown even in Sandage's *The Hubble Atlas of Galaxies* (p. 41). Alar Toomre described the change of paradigm, writing that for many researchers of the 1970s, "NGC 520 has instead seemed a pairwise galaxy merger well underway ... whereas M82 has instead been exhibiting serious signs of indigestion brought on by eating too much interstellar gas from the outer disk of M81 during a fairly recent (and direct-sense) close passage that seems also to have caused the unusually wide-open pair of beautiful outer spiral arms in M81 itself!"[16] (Plates 7.1 and 11.1).

An interesting story is that of the Carnegie Observatories astronomer François Schweizer: "In the middle of the summer of 1972, Ivan King came back from a meeting and handed me a preprint of the Toomre & Toomre (1972) paper on Galactic Bridges and Tails. In reading this preprint, I instantly understood that they were opening a first window on galaxy evolution, and I immediately turned much more to Arp's *Atlas of Peculiar Galaxies* to study the subject and prepare a research proposal for my job applications in 1973. It is this proposal that got me a Carnegie Fellowship (1974–75) and launched me into researching whether ellipticals could originate from mergers of spirals, as TT72 suggested. The rest is history."[17,18] Arp's atlas inspired the work of François Schweizer and the Space Telescope Science Institute astronomer Brad Whitmore's long research program on interacting galaxies. They demonstrated that collisions and mergers have shaped galaxies and determined their stellar and gaseous content in ways Arp would not have expected.

The digital age has brought with it a massive number of images, which has changed the course of the impact of atlases. The latest and most unusual galaxy project is the Galaxy Zoo, a wonderful example of the emergence of "big data" science and a new way of involving a very broad community in science projects. Huge samples are used to nail down better the transformative processes that galaxies undergo during their lifecycle. The Indian astronomer Preethi Nair of the University of Alabama worked at classifying several thousands of galaxies from the Sloan Digital Sky Survey. She went through many of the atlases and wrote: "The version I used as a template for my classification was *The Carnegie Atlas of Galaxies*. The book sat on my desk for two years. One reason I used it more often was because it was available on-line so I could refer to it from any location."[19]

These examples of reactions and inspirations for programs of researchers are only a few among many others. In summary, atlases of galaxies with carefully organized images and sequences of images are a sort of "shopping window" on what the universe is offering as products of complex physical processes with long evolutionary histories. The atlases described in Chapter 10 differed enough from each other to provide contrasting and incremental views on the extremely rich world of galaxy shapes. The diversity of programs and projects that emerged from looking at and working with atlases is a powerful testimony to the fact that images, especially in their systematic presentation within these atlases, triggered the generation of new knowledge.

---

[16] A. Toomre, private e-mail to the author (October 2014).   [17] Private e-mail note to the author (September 2014).
[18] A. Toomre and J. Toomre, Galactic Bridges and Tails, *The Astrophysical Journal*, 1972, Vol. 178, pp. 623–666.
[19] Preethi Nair, private e-mail to the author (October 2014).

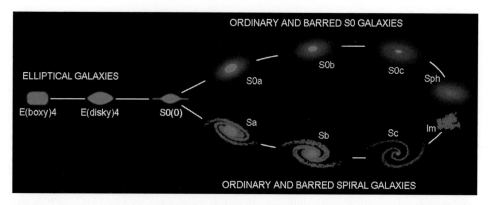

Fig. 11.1 Revised S0 branch. From *The Astrophysical Journal Supplement.* © AAS. Reproduced with permission.

And now for a a trickier question: Can one predict shapes/objects using the classification schemes? A fascinating recent development in the physical morphology of galaxies is that the so-called lenticulars, S0, once considered to be transitional (Chapters 6 and 9) between ellipticals and spirals (Sa) may not be transitional at all. Sidney van den Bergh had already suspected this in 1998.[20] The American astronomer John Kormendy and the German, Ralf Bender, have proposed that they should be added as a new separate branch parallel to spirals. This challenge to the Hubble tuning fork has received solid support from recent observations and numerical simulations.[21] The new sequence would be a natural home of an important class of galaxies that has been a "troublesome kid" for decades. "The revision to the tuning fork represents the coming to fruition of quantitative and interpretative galaxy morphology."[22] This is also a dramatic demonstration that classification schemes of galaxies, as illustrated by atlases, can lead to completely new perspectives. In 2012, Kormendy and Bender proposed a major revision of the parallel-sequence morphological classification of galaxies. They added a parallel sequence with lenticulars having their own S0a, S0b, S0c and Sph suite to mark the bulge-to-total ratio diminishing from left to right (Fig. 11.1).

Book Reviews of the Atlases: What do they Say?

Book reviews offer interesting windows into the contemporary research and academic context of a given discipline at the time of publication. All atlases were objects of book reviews, but in varying numbers; these reviews were generally published in well-known astronomy or general science magazines (Table 11.1). Reviews that appeared shortly after the publication of the various atlases are also revealing of their reception and of their expected impact.

---

[20] S. van den Bergh, *Galaxy Morphology and Classification*, Cambridge: Cambridge University Press, 1998.

[21] J. Kormendy and R. Bender, A Revised Parallel-Sequence Morphological Classification of Galaxies: Structure and Formation of S0 and Spheroidal Galaxies, *Astrophysical Journal Supplement Series*, 2012, Vol. 198.

[22] R. J. Buta, Galaxy Morphology, In *Planets, Stars and Stellar Systems*, Vol. 6: *Extragalactic Astronomy and Cosmology*, Dordrecht: Springer, 2013, p. 83.

Today, they make fascinating and revealing reading. According to their general remit, these book reviews highlighted the innovative features of the atlases, described their format and presentation, and estimated their potential impact on the discipline. The emphasis by the reviewer was often about the quality of the publication and format, as expected for major reference books such as atlases. The comments on the aesthetic value of the atlases were often surprising and entertaining. Let us pick a few of the most noticeable perspectives.

On *The Hubble Atlas of Galaxies* (1961), American astrophysicist George Field wrote in glowing terms in *The American Scientist*: " . . . at $10.00, it is a bargain you are not likely to find again in your lifetime." Highlighting the breathtaking illustrations, he added: "An artistic achievement, this volume is also a scholarly work of major proportions."[23] In the *Journal of the Royal Astronomical Society of Canada*, Sidney van den Bergh praised "the finest collection of photographs of external galaxies" ever published, but he did not refrain from a minor criticism: for disk galaxies, the S0 and Sa types may not correspond to a "physically distinct type of object" and instead appear to have become "depository for systems having widely differing physical characteristics."[24] There appeared to be a missing link between ellipticals and disk galaxies: S0 were thought to be the intermediary step. Van den Bergh felt the class was a "hotchpotch" box. Considering the more recent proposal of Kormendy and Bender, just mentioned above, that lenticulars were a fully separate morphological class, this early comment from van den Bergh was visionary.

Writing an extensive review for the prestigious scientific weekly *Science*, the American astronomer Frank Edmondson quoted generously from Sandage's introductory text to the atlas:[25] "*The Hubble Atlas of Galaxies* is a major contribution to the observational study of cosmology." He even carried out a detailed account of the observers who had obtained the photographs used in the atlas. The 186 photographs in all of the 176 different galaxies contained in the atlas were distributed as follows: Hubble took 72, Sandage, 63, Milton Humason, 24; and seven other astronomers took the rest. The use of halftones and the very high quality of the material was praised as stimulating an "esthetic appreciation for fine astronomical photographs."

Commenting in *Astronomy* magazine on the gigantic size of the *Atlas of Galaxies Useful For Measuring the Cosmological Distance Scale* (1988), Jeff Kanipe wrote in a humorous quote that the new giant atlas, "the biggest galaxy book ever," could have legs added and become a coffee table! Walter S. Houston of the magazine *Sky and Telescope* highlighted in particular the high quality of the images and re-stated the purpose set out by the authors, Sandage and Bedke: "The programmers of the HST can use these prints to pick out regions where Cepheids should be easiest to find." The Cepheids of dozens of nearby galaxies were indeed used to nail down the value of the Hubble constant, i.e. the rate of the expansion of the universe, with extreme accuracy. The goal was achieved following a multi-year observing program with a key project using the HST led by the Canadian astronomer Wendy

---

[23] G. Field, Book Review: The Hubble Atlas of Galaxies, *American Scientist*, 1962, Vol. 50, p. 212A.
[24] S. van den Bergh, Review of Publications: The Hubble Atlas of Galaxies, *Journal of the Royal Astronomical Society of Canada*, 1962, Vol. 56, pp. 29–30.
[25] F. Edmondson, Reconnaissance of Outer Space, *Science*, 1961, Vol. 134, p. 464.

Freedman in the 1990s. The result: the Hubble constant is 70 km/s per megaparsec (or about 20 cm/s per light-year).

The Harvard Astronomer Margaret Geller, writing about the monumental *The Carnegie Atlas of Galaxies* (1994), was extraordinarily eloquent about the two-volume atlas, the final and greatest work by Sandage and Bedke: "This atlas is a work of art. Nowhere are the grandeur and aesthetic appeal of the cosmos more elegantly displayed.... *The Carnegie Atlas* should be in art museum shops along with other collections of famous photographs."[26] The British astronomer Gerry Gilmore wrote: "I have not previously reviewed a scientific or technical book using remotely as many superlatives as I have here.... It is a truly outstanding contribution to astronomy." Then qualifying his enthusiasm in more sober terms: "The clear and objective exposition transforms this atlas and classification scheme from a transient, personal, and possibly somewhat subjective phenomenon to a basic and enduring tool for scientific research. It ensures its continuing utility for many years to come, and makes this atlas a major milestone on the pathway to enhanced scientific knowledge."[27] With his educationalist hat, Gilmore gets close to preaching: "Those young scientists need this atlas. So do we old ones.... The young researchers to whom the atlas is addressed are being offered a helping hand up, to stand on the shoulders of a giant." But the atlas might have been a big step for the newcomer. Geller suggested that a separate informal guide for young people and their teachers would have been a valuable companion to the atlas.

In the popular French *Ciel et Terre*, Belgian astronomer René Dejaiffe notes that *The de Vaucouleurs Atlas of Galaxies* (2007) is "... the first atlas based on digital detectors instead of photographic plates. The images of the 523 galaxies of the atlas also come from a combination of ground-based and space-based telescopes. The extensive discussion and interpretations presented make this atlas a new reference for galaxy work."[28] Indeed, Buta et al.'s recent atlas had finally done justice to the very fine and precise classification scheme proposed by de Vaucouleurs more than half a century previously (Chapter 10). Stewart Moore, in a review for the *Journal of the British Astronomical Association*, recalls the limitations of the Hubble tuning-fork-shaped classification system and its challenge to deal with the junction between the ellipticals and the spirals, again the headache of the lenticular (S0) galaxies, as noted by Sidney van den Bergh in his review of *The Hubble Atlas of Galaxies* of 45 years earlier.[29] There have been many indications that things in galaxy morphology were subtle and complicated. Early in the 1950s, de Vaucouleurs had introduced a three-dimensional diagram creating a classification "volume" that dealt more systematically and rigorously with the secondary features of galaxy structure (Chapter 10, see discussion on the *The de Vaucouleurs Atlas of Galaxies*). Moore went on to discuss the many indications that things in the galaxy world were subtle. "The long axis contained the main classes of

---

[26] M. Geller, Book Review, *Science*, 1995, Vol. 268, p. 1214.
[27] G. Gilmore, Book Review, *The Observatory*, 1995, Vol. 115, No. 1128, p. 278.
[28] R. Dejaiffe, Book Review, *Ciel et Terre*, 2007, Vol. 123, p. 121.
[29] S. L. Moore, Book Review: The de Vaucouleurs Atlas of Galaxies, *Journal of the British Astronomical Association*, 2007, Vol. 117, no. 4, p. 211.

galaxies while a cross-section at each class described families within that class." He noted that the authors of the atlas had modified and improved de Vaucouleurs' system to include new findings with space-based observatories and infrared imaging.

Authors of atlases and architects of the classification systems themselves commented on other atlases. We read in the preceding chapter praise by Allan Sandage for Wray's *The Color Atlas of Galaxies*. Sidney van den Bergh also liked the atlas and recommended it as follows: "For an excellent introduction to the beauty and variety of galaxy morphology, the reader is referred to *The Color Atlas of Galaxies*".[30] The blunt and critical opinions of Halton Arp on Sandage's *The Hubble Atlas of Galaxies* have already been discussed (Chapter 10). Gérard de Vaucouleurs wrote a detailed review, in Spanish, of José Luis Sérsic's *Atlas de galaxias australes*, emphasizing the importance of the photometric maps and data, which the atlas was the first to include.[31] Vera Rubin found the photographs of the *RSA* "exquisite," and being a very practical researcher, added: "In among all the goodies is Appendix A: a table of additional galaxies," which expanded on the original Shapley–Ames catalogue.[32]

Often written by well-known astronomers, book reviews added an aura of authority that sent the appropriate signals to librarians to make a safe purchase, or to the individual keen to acquire the work, especially graduate students who wanted a guaranteed intellectual return for their investment. As mentioned before, astronomy is a discipline that enjoys a world-wide community of amateurs. They form a strong group of readers and users of atlases, as demonstrated by the numerous reviews of atlases that were written by both professional and amateur astronomers and published in astronomy magazines. More recently published, *The Arp Atlas of Peculiar Galaxies* has amateurs as its prime target audience.[33]

A few atlases received criticisms: Wray's *The Color Atlas of Galaxies*, for the very poor quality of reproduction of the images (the criticism was aimed at the publisher), or Hickson's *Atlas of Compact Groups of Galaxies* for which Irish astronomer David Andrews questioned the physical relevance of the groups and the selection of objects, and was hoping for an improved revised edition.[34] The British astronomer Jeremy Allington-Smith wrote that Hickson had assembled "an objective catalogue," but "as a coffee-table book, it must be judged a failure since its black-and-white pictures are too humdrum, despite the inclusion of every astronomer's favorite pinup – the original Stephan's quintet."[35] While not being a bedtime reading book, Hickson's atlas – the first to employ CCD imaging – enjoyed the largest number of book reviews.

The recent *The de Vaucouleurs Atlas of Galaxies* has also received a mixed set of reviews for its "dull presentation" aimed at the specialist: "no pretty colour images," and "not for beginners." In my opinion, these statements are rather unfair, as the reviewers overlooked

[30] S. van den Bergh, *Galaxy Morphology and Classification*, Cambridge: Cambridge University Press, 1998.
[31] G. de Vaucouleurs, Algunos comentarios sobre fotometria de galaxias y el "Atlas de galaxies australes" de Jose Lui Sérsic, *Revista Astronomica*, 1969, Vol. XLI, pp. 27–34.
[32] V. Rubin, Book Review: A Revised Shapley–Ames Catalogue of Bright Galaxies, *Sky and Telescope*, 1982, Vol. 63, p. 478.
[33] J. Kanipe and D. Webb, *The Arp Atlas of Peculiar Galaxies*, Richmond: Willmann-Bell, 2007.
[34] A. D. Andrew, Book Review: Atlas of Compact Groups of Galaxies, *Irish Astronomical Journal*, 1995, Vol. 22, p. 226.
[35] J. R. Allington-Smith, Book Review: Atlas of Compact Groups of Galaxies, *The Observatory*, 1995, Vol. 115, p. 347.

the advantages of the logarithmic, sky-subtracted digital images, such as revealing details of both the inner and outer structure and also the homogeneity of the illustrations. Furthermore, the images in *The de Vaucouleurs Atlas of Galaxies* are no less appealing than those of *The Hubble Atlas of Galaxies* or of *The Carnegie Atlas of Galaxies*. Nevertheless, from these reviewers it is clear that there is an expectation for atlases to be well produced and visually attractive, almost like art books.

Overall, the majority of galaxy atlases garnered overwhelmingly positive and fair reviews, albeit with a few cases of overenthusiastic pronouncements on the potential usage. The reviews most certainly helped to stimulate the market for the atlases. As opposed to the great scientific atlases of the nineteenth century, which had to be sold through advanced sub-scriptions to sustain their outlandish production costs, atlases of the galaxies went straight to the market, like normal books.

Finally, it is worth noting that the number of citations to the atlas (i.e. its apparent use) does not necessarily correlate with the number of book reviews received. For example, *Atlas de galaxias australes* had only a single review by de Vaucouleurs, as part of a longer commentary published in an obscure Argentine journal on the Sérsic formula describing galaxy structure. And yet it is the most cited atlas, alongside Sandage's *The Hubble Atlas of Galaxies*. Similarly Arp's atlas, published in a very specialized research journal, did not receive many reviews, perhaps on account of the emphasis by the controversial author on "rogue" galaxies and his strong contrarian approach to galaxy classification.

The Scholarly Driving Forces for Galaxy Atlases

Lorraine Daston and Peter Galison have written in eloquent and illuminating ways about the many purposes and driving forces of scientific atlases and of their role in the research and learning process. "Atlases are systematic compilations of working objects. They are the dictionaries of the sciences of the eye. For initiates and neophytes alike, the atlas trains the eye to pick out certain kinds of objects as exemplary and to regard them in a certain way.... The atlases drill the eye of the beginner and refresh the eye of the old hand."[36] Atlases of galaxies fit this broad description in their design, production and usage.

The authors of atlases and their publishers carefully put together selections of "stan-dardized objects" for researchers and students. Hence, as shown through examples in this chapter, atlases have been tools for researchers (by assembling archetypes) and for learners/students (by showing well-identified exemplars) to view, explore and understand the morphological diversity of galaxies. They have been "working objects" in the full sense.

Thus equipped, it becomes possible for the researcher or the student to relate shapes and morphology to various cosmic phenomena; for example:

- environmental density, which affects the interaction between galaxies and their merger history;

---

[36] L. Daston and P. Galison, *Objectivity*, New York: Zone Books, 2007, pp. 22–23.

- internal perturbations (bars) and secular evolution, which determine the strength of the spiral arms and bars: or the presence of other resonances, such as ovals, rings and lenses;
- gas accretion, which may sustain the nuclear activity and correlate with the presence of a supermassive black hole at the galaxy center;
- the mix of structures, which indicates the instabilities or past accretion events: disks with an exponential radial profile, bulge and nucleus, and dust lanes; and
- processes leading to massive star formation and differences between objects, even of similar classes.

Galaxy atlases have been used beyond observing programs. They have provided sets of "working objects" for theoreticians and "simulators" as highlighted in the early and pioneering work of Alar and Juri Toomre. Galaxy morphology gives clues on formation and evolution – there is order, as multiple chaotic processes do not dominate galaxy shapes as is the case for nebulae. Deviations such as "tails" or outer shells can in the end be understood as the result of tidal effects, and bars as dynamical instabilities in a large assembly of stars, orbiting with a slightly asymmetric gravitational potential. Sorting galaxies by morphology has been possible and meaningful, just as it is possible to sequence stars using their spectra. While schemes of galaxy classification have been driven by optical observations, the arrival and recent use of infrared imaging has reinforced the robustness of the scheme proposed by Hubble as early as 1927 (and improved by Sandage in 1961), refined by de Vaucouleurs in 1959, and reorganized by Kormendy and Bender in 2012.

Galaxy classification is undergoing a revival.[37] It has become a vibrant area of citizen science as the digital age gives access to huge samples of galaxies. Astronomy is entering a new era.[38] Furthermore, the HST, looking backward in time, is pushing galaxy classification to higher redshifts and challenging our local universe classification schemes (see Plate 11.2). Galaxies look different, based on their star formation rate and active interaction history. "Zooites" is the name given to the people who participate in the Galaxy Zoo, a vibrant citizen science project. More than 50 million galaxy classifications were received in the first year from almost 150,000 people. The Zooites of Galaxy Zoo 2, who work with ultra-deep HST deep images, have noticed a greater number of irregular galaxies in Hubble deep fields compared to previous studies. A large number of galaxies do not fit the traditional profile. They have come up with new classes, for example "pea galaxies," which are compact galaxies undergoing very intense bursts of star formation (Fig. 0.9). But this push to include weirder objects will allow the scientific tradition of atlases to continue to be a repository of knowledge and a vehicle for identifying new classes of objects.

Some Conclusions about Galaxy Atlases

What conclusions can we draw from this survey and analysis of galaxy atlases? The main points may be summarized as follows:

---

[37] See the mid-infrared atlas based on the Spitzer Survey of Stellar Structure in Galaxies by Ronald Buta et al., accessible at http://kudzu.astr.ua.edu/s4g-morphology/s4g-morphology.html and discussed in the appendix.

[38] See K. W. Willett, C. J. Lintott, S. P. Bamford et al., Galaxy Zoo 2: Detailed Morphological Classification for 304,122 Galaxies from the Sloan Digital Sky Survey, *Monthly Notices of the Royal Astronomical Society*, 2013, Vol. 435, pp. 2835–2860.

1. Images of galaxies have been a powerful discovery tool for understanding the nature of galaxies and for inferring the processes of their formation and evolution.
2. Morphology has been, and continues to be, a most powerful parameter for sorting the categories of galaxies.
3. When classification schemes appear to be in opposition – even contradictory, such as those of Hubble–Sandage versus Arp's – they are actually describing galaxies at different stages.
4. Atlases based on optical imaging have been powerful tools to understand galaxies at all wavelengths; imaging in the infrared (3.6 micron) appears promising for classification.
5. The Shapley–Ames catalogue of 1932 (and its revision by Sandage and Tammann – the *RSA*) played a key role (by defining samples of objects) in galaxy research programs of the 1980s and 1990s.
6. The number of citations of atlases are uncorrelated with the number of reviews at the time of publication.
7. The quantitative (and unifying) approach of José Luis Sérsic contributed to his less-well-known atlas being one of the most cited galaxy atlases.
8. With digital archives and powerful tools of enquiry, galaxy atlases are bound to undergo significant changes.

Finally, galaxy classification has been at the root of many cosmological studies, for example the fraction of spirals as a function of redshifts, or the presence of a bar in relation to the age of the universe. Studies of distant and young galaxies are rooted in the simplest elements of galaxy classification from nearby objects. "Remarkably galaxy evolution models across the history of the universe are often based first and foremost on morphological signatures first identified 150 years ago!"[39]

---

[39] P. Martin in an e-mail note to the author (February 2015).

Conclusion

This was the astronomical observing experience at its best – a dark, quiet dome, a silently moving monster telescope, and mastery of the dangerous Newtonian platform, all in the interest of collecting data on a problem of transcendental significance.

Allan Sandage[1]

What is the Significance of Galaxy Images?

Allan Sandage (see Fig. 10.3) was a giant of twentieth-century astronomy. He spearheaded the advancement of our knowledge in key areas of astrophysics. He contributed fundamentally to our understanding of the universe by decrypting stellar evolution. Sandage made fundamental contributions to constructing and interpreting Hertzsprung–Russell (H–R) diagrams of star clusters that display stars with their luminosity as a function of effective temperature. As non-homomorphic representations, such H–R diagrams have been crucial in revealing that stars evolved not by sliding down the "main sequence," but by drifting on well-defined tracks across the diagram, changing in luminosity and effective temperature throughout their lifecycle, at rates depending on the initial mass of the star. Sandage delved into the nature of mysterious radio sources, discovering the first quasar, and helped to refine the extragalactic distance scale. He measured the properties of expanding spacetime.[2] Sandage's dream was nothing less than "decoding cosmic evolution."

Throughout his whole career Sandage accomplished most of his work with indomitable energy and passion by taking, analyzing and interpreting images. He pushed for building one of the best imaging telescopes, the 2.5-m Irénée du Pont telescope at Las Campanas, Chile, with its large field of view. For him, images were full of secrets just waiting to be deciphered by the persistent and scrutinizing mind, which could reveal what he expected to be the deeper nature of our universe. A true artist of the scientific image, he shared his work and strong vision in several magnificent ways. As main author of four of the great galaxy

---

[1] A. Sandage, *Centennial History of the Carnegie Institution of Washington, Volume 1: The Mount Wilson Observatory*, Cambridge: Cambridge University Press, 2004, p. 178.
[2] D. Lynden-Bell and F. Schweizer, Allan Rex Sandage, 18 June 1926–13 November 2010, *Biographical Memoirs of Fellows of the Royal Society*, 2012 (arXiv:1111.5646).

atlases, Sandage passed on an inspiring and lasting legacy. Few astronomers have used and demonstrated in more vivid ways the power of images for scientific discovery.

Science seeks order in the natural world of "things." As I have shown in this book, images have played an essential role in discovering order in the sidereal world, in unveiling galaxies as "island-universes" and in establishing the astounding structure and dynamics of the expanding universe. Allan Sandage was certainly a master and a leading explorer in this complex process. Along with Gérard de Vaucouleurs and the iconoclastic Halton "Chip" Arp, he drew for us some of the best roadmaps to the universe.

Images as Roadmaps to the Universe

But what about all these images? What has been their role on this astounding path of discovery? I surmise that images of galaxies crystallize our cosmic knowledge along four cognitive dimensions. "Without a visual first-hand impression we are literally fumbling in the dark, but once we have an image, we rarely look back at it, as it becomes a 'simplicity'."[3] Images of galaxies help us (i) to grasp the vast physical scale of the universe, (ii) to climb the ladder of cosmic complexity, (iii) to sharpen our aesthetic and epistemological sense in positioning ourselves in nature at large and, finally, (iv) to drive the design and building of increasingly powerful instruments, while initiating new research programs.

First, images are most powerful in representing at a human scale objects of fantastically different spatial scales. For example, objects as different as superclusters of galaxies that embrace 10 million light-years of length in space, animals of human size, plant cells of a few micrometers or silicon atoms on the surface of a crystal at the nanometer scale can all be displayed in a page of a book or a scientific atlas or on a computer screen. At the upper rungs of the cosmic ladder, images help us visualize the immense magnitude of the universe: from successive images going from the Earth–Moon tandem, to our solar system planetary world, to nearby star clusters like the Pleiades, to the larger globular clusters, to the Milky Way and its Magellanic Cloud satellites, to the Local Group of galaxies, to the Virgo cluster of galaxies, and further still. Hence, we move from scales of light-seconds to millions and billions of light-years. Images help us to assemble a cognitive chart of the universe and of its components and to grasp their wildly diverging physical scales.

Secondly, images of astronomical objects and their history reveal the amazing diversity and range of structures of celestial objects. "The study of size, shape, brightness, central concentration, degrees of mottling, rotation, change, and movement, as well as questions of morphology, identity, classification, and evolution, are current to contemporary nebular and extragalactic astronomy, just as they were in the nineteenth century."[4] Seeing and viewing while scripting notes was an initial step in exploring celestial features. However, even the best and most detailed written notes did not suffice. It would have been challenging for anyone to take the observing notes of a skillful nineteenth-century observer and try to

---

[3] Lars Lindberg Christensen, e-mail to the author of November 2016.
[4] O. W. Nasim, *Observing by Hand: Sketching the Nebulae in the Nineteenth Century*, Chicago: University of Chicago Press, 2013, p. 233.

reconstruct even an approximate picture just from his/her comments and descriptors. Sketching of nebulae, in support of the act of viewing and observing, illustrated their complexities. Then twentieth-century technologies totally transformed these processes. Astrophotography, aperture synthesis using radio waves and X-ray imaging revolutionized astronomical imaging; they provided us with tools to examine cosmic objects in the finest details over the complete electromagnetic spectrum in previously undreamt-of ways. These latter steps were essential to understand the nature of cosmic objects and to reveal the complex underlying physical processes.

Using a battery of telescopes and sensors, we became able to monitor objects over time, either by following single objects or by comparing vast ensembles of cosmic objects of the same class. We established that astronomical timescales are to be measured in thousands, millions or billions of years. Variations of brightness of many astronomical objects over time have turned out to be incredibly rich, even over relatively short timescales. Powerful imaging techniques have revealed significant variations in some objects over time on the hourly and even daily scale. Some phenomena, such as pulsars, fluctuate on millisecond timescales, which is amazingly short for astronomical objects. The firmament of fixed stars has been blown apart.

Thirdly, images of galaxies may not be as spectacular as those of solar system objects taken by interplanetary space probes. They may not offer the diversity of forms or colours of butterflies, birds, or palm trees, and far less than many of the natural objects illustrated so spectacularly by Ernst Haeckel in *Kunstformen der Natur*, a book of lithographic and halftone prints. We may also not see the finest morphological or kinematic details such as those of the atmosphere of Jupiter, the rings of Saturn or the icy pingos of Pluto. Nor do we see in the average galaxy the explosive range of colours of galactic nebulae (unless false colours are used to highlight features). Despite these relative pictorial shortcomings, galaxies at first sight do display order and stability on the grandest scale of nature. And at the other extreme, they unveil the ultimate cosmic cataclysm, the grandiose chaos of colliding and merging galaxies. Reconstructed by computer simulations, real images of merging galaxies make us realize our smallness: how fleeting and contingent our existence is, how small our world, itself sailing around an inconspicuous dwarf star carried around with billions of others in the giant Milky Way merry-go-round.

Fourthly, many impressive telescopes have been built and continue to be built, each generation more powerful and an improvement on the previous one. The first great reflectors of the twentieth century were designed in great part to observe and explore the nature of "nebulae," when most of them became "galaxies." Creative opticians and builders drove on relentlessly for better and deeper imaging capabilities: in succession, we had William Herschel's 40-ft reflector, William Parsons' 6-ft Leviathan, the Crossley 36-inch, the Ritchey–Hale–Pease's Mount Wilson 60- and 100-inch, the Palomar 5-m, Irénée du Pont 2.5-m, and finally the Hubble Space Telescope, the latter being one of the most expensive and productive scientific machines ever built. The 6.5-m James Webb Space Telescope, due to be launched in 2018, has as one of its core scientific missions the imaging of the first galaxies that formed more than 12 billion years ago. On the ground, following the dozen or so 8–10-meter giants (Subaru Telescope, twin Gemini telescopes, the Very Large Telescope, the

Keck twin telescopes and a few others), three new mammoth telescopes of 25–40 m in size are under construction. The Atacama Large Millimeter Array (ALMA), the great submillimeter wavelength interferometer on the altiplano of northern Chile, is providing unique views of young planetary systems surrounding nearby stars and of molecular clouds in distant forming galaxies. Their imaging capabilities would have been inconceivable only 50 years ago.

The Digital Universe

We are at the dawn of a new age in astronomical imaging. Soon, telescopes will generate terabytes of imaging data, day and night. What will happen with the widely used digital imaging and computer processing of astronomical images? What is the impact of the computer age? There is no doubt that the digital age is changing the role and use of atlases of galaxies, and of scientific imaging. This is evident from just flipping the pages of the weekly issues of the journals *Science* and *Nature*, or browsing the daily posting of new images on the NASA website, Astronomical Picture of the Day.[5] Modern computer-based catalogues and sophisticated query tools have become part of everyday research life. A powerful synergy is growing between the new tools and the large synoptic surveys of the sky from the ground and space. Multiwavelength views have superseded photographic plates. By cleverly combining images of a given object obtained at different wavelengths and stacking them like a sophisticated pack of virtual cards, researchers now work with multiwavelength data cubes and multidimensional data entities.

With all these electronic advances, atlases are also entering a new age.[6] With the powerful panoramic detectors, we have moved from producing analogue images to digital ones. This emphasizes the key difference between analogue and digital. Photographic plates, the best analogue photographic detectors, were non-linear in their response to light: that is, an increase of light by a factor of four did not translate into a factor of four of darkening of the emulsion; the chemical reaction was instead some complex response function that was tricky to calibrate and required time and training on the part of the observer. With digital electronic detectors like CCDs, the response is linear; ten times as many photons result in ten times more electrons. Furthermore, digital detectors or imagers generate a quantifiable number ready for the mathematical operations of computer image processing. The handling of images has become a sort of mind aerobatics, and many astronomers have become extremely adept at this. Several user-friendly tools and applications have been created and shared, enabling complex image manipulations to be carried out by amateur astronomers and students, not only professionals.

What will the future galaxy atlases be? It is now possible to do things that were impossible with printed atlases, while still serving the same purposes and in a far more versatile way. We have the ability to make images with extremely large numbers of pixels covering

[5] http://apod.nasa.gov

[6] I am most grateful to Lars Linberg Christensen (European Southern Observatory) and Anton Koekemoer (Space Telescope Science Institute) for sharing their perspectives on the role of atlases in the digital age.

a much higher dynamic range of brightness. This provides a far greater cognitive power in viewing the overall scene or for zooming in on details. With a few keyboard clicks, we are able to vary image parameters such as the filter or the wavelength range we wish to use to examine a field, bringing out the underlying physics. Computer-aided, one can also skim through thousands of images in a matter of minutes, arrange appropriate sequences and tune into the desired wavelength of interest. These manipulations can reveal subtle commonalities and differences between images.

Astronomers have assembled a colossal treasure: millions of images of the sky at multiple wavelengths and multiple tools to analyze them. For example, the Virtual Observatory is a collection of astronomical data centers. Among its many tasks and projects, it offers software systems and powerful image processing capabilities to researchers. By appropriately mastering and enabling tools of artificial intelligence, astronomers have new ways of browsing through archives and are able to combine images obtained, for the same object, in the X-ray, the visible and the radio domains. It is important to note some of the challenges of multiwavelength imaging. When images come from different telescopes and instruments spanning the electromagnetic spectrum, how do we ensure proper alignment of the images? If we correct the images for geometric distortions, are we compromising the measurements? The alignment of features in different spectral regions does not necessarily mean that they are the same physical object.[7] The tools offered by the Virtual Observatory help to address these issues and make automated multispectral imaging analysis, interpretation and classification rewarding. Perhaps, one day, the galaxy researcher will be able to construct his or her atlas on demand.

Galaxy Images for Everyone

The best example of the new age is the Hubble Legacy Archive.[8] This is one of mankind's finest modern atlases and it is growing. The people at the Space Telescope Science Institute (STScI/NASA), the Space Telescope European Coordinating Facility (ST-ECF/ESA) and the Canadian Astronomy Data Centre (CADC/NRC/CSA) have also made considerable efforts to visualize the Hubble Space Telescope data as well and as efficiently as possible. The data in the thumbnails and "interactive displays" are uniformly treated and provide an excellent way for scientists to get a first-hand impression of the data.

Innumerable things remain to be discovered from archival material, as old photographic plate collections are digitized and loaded into modern archives. It is crucial that this old observatory material is protected and not lost. The colossal effort of cleaning and digitizing the 500,000 photographic plates of the Harvard College Observatory plate stacks deserves particular praise.

---

[7] Glenn Mackie, *The Multiwavelength Atlas of Galaxies*, Cambridge: Cambridge University Press, 2011.
[8] https://hla.stsci.edu/hlaview.html

Several observatory outreach departments are deploying significant efforts to produce ethically correct, impressive colour imagery from the raw telescopic data.[9] Started many years ago as Pictures of the Week projects for both the European Southern Observatory and Hubble Space Telescope, this developing archive assembles images from existing databases or published material. Images are tagged with various metadata (coordinates, orientation, wavebands, etc.) providing context that may be useful for both laypeople and experts. Often representing the best images of the objects in question, this project constitutes a slowly accumulating archive, close to an electronic atlas. AstroPix is another project combining all available images (across all wavelengths) into one searchable interface.[10] Finally, images are also uploaded to Wikipedia that may arguably be the biggest atlas in the making.

What a long and fascinating road we have traveled since Al-Sufi's description of the "little cloud" in Andromeda in AD 964. Let us recall that the pioneering observers of the eighteenth and nineteenth centuries had no idea what "nebulae" were: a sidereal fluid, some interstellar fog, mysterious clouds, other solar systems in formation, unresolved clusters of stars, or distant stellar systems? While we now know and understand the basics of "island-universes," it is humbling to realize that this epiphany occurred less than 100 years ago and that the disconcerting puzzles of dark matter and dark energy have emerged as the new frontiers of sidereal exploration. Let us close with English philosopher Francis Bacon (1561–1626) whose wisdom may outlast galaxies. "We may, therefore, well hope that many excellent and useful matters are yet treasured up in the bosom of nature, bearing no relation or analogy to our actual discoveries, but out of the common track of our imagination, and still undiscovered; and which will doubtless be brought to light in the course and lapse of years, as the others have been before them; but in the way we now point out, they may rapidly and at once be both represented and anticipated."[11] Far from being powerless, our minds make us capable of exploring the "bosom" of nature. In employing an amazing arsenal of imaging tools and techniques, humans continue to dream of exploring the universe in its entirety, with a deep desire to decipher its significance and their own place within it.

---

[9] See www.eso.org/public/images/archive/category/galaxies/ and www.spacetelescope.org/images/archive/category/galaxies/ where thousands of galaxy images are available.

[10] See AstroPix at astropix.ipac.caltech.edu

[11] Francis Bacon, *Aphorism 109*, Novum Organum, 1620.

Appendix

Atlases for Special Usages

This appendix is a complement to Chapter 10 where the main galaxy atlases published since 1961 are presented. These atlases were based on photographic material of images of galaxies at optical wavelengths. In addition to these, there have been a number of specialized atlases, which are targeted at experts in the field. These works were based on a selected number of galaxies. Often they were published in professional journals.[1] There have been very few atlases of galaxies at wavelengths other than optical and infrared, some of which are reviewed below.

Sidney van den Bergh's monograph on *Galaxy Morphology and Classification* occupies a special place.[2] It is not an atlas in itself but it addresses the fundamental questions about morphology being the most common tool to classify galaxies. It is also a fine introduction to the galaxy classification systems of Hubble and Sandage, de Vaucouleurs, Morgan (Yerkes System), Elmegreen and Elmegreen (the arm classification system) and that of the author himself – the David Dunlap Observatory (DDO) system. The book is concise and deals very effectively with the challenges of galaxy classification based on optical imaging. It provides a critical overview of the successes, limitations and failures of the various schemes.

William Morgan wrote in 1958, "the value of a system of classification depends on its usefulness." Let us go over a few of those specialized atlases to give a flavour of their scope and purpose, and ultimate usefulness.

Radio Mapping of Galaxies

Arnold Rots (1979) put together the *Atlas of 21 cm HI line Profiles for 61 Galaxies of Large Angular Size*, based on radio observations of galaxies conducted with the 300-ft (91-m) National Radio Astronomy Observatory telescope at Green Bank, West Virginia.[3]

---

[1] A fine example is B. T. Lynds, An Atlas of Dust and HII Regions in Galaxies, *Astrophysical Journal Supplement Series*, 1974, Vol. 28, pp. 391–306. The plates of 41 galaxies include photographs accompanied by sketches where HII regions and the dust band are shown.

[2] S. van den Bergh, *Galaxy Morphology and Classification*, Cambridge: Cambridge University Press, 1998.

[3] A. H. Rots, *An Atlas of 21 cm HI Line Profiles of 61 Galaxies of Large Angular Size*, Green Bank: National Radio Astronomy Observatory, 1979.

The sample included all spirals and irregulars north of declination −19° with a large angular diameter. The galaxies chosen by Rots had Holmberg radii between 9 and 36 arcmin; this is the radius where the galaxy has a surface brightness of 26.5 blue magnitude per square arc-second. The observations were presented in the form of global radial profiles and integrated neutral hydrogen contour maps, published separately.[4] The purpose of the atlas, which had no photographs, was to display the large-scale coherent structure of neutral hydrogen. The contour maps were printed at the same scale as the Palomar Sky Survey (67.4"/mm) so that transparencies made from these figures could be directly overlaid on Sky Survey prints.

A Census of Galaxy Families

Galaxies rarely come isolated. Large assemblies of several tens or even thousands of galaxies over volumes of a few megaparsecs have been known for some time. Fritz Zwicky had systematically observed clusters of galaxies; he produced a massive catalogue of 9,134 galaxy clusters in several volumes. Later, the American astronomer George O. Abell (1923–1983) used the Palomar Sky Survey to catalogue 2,712 clusters (the Northern Survey in 1958) and a further 1,361 clusters (the Southern Sky Survey in 1989) from the SRC–ESO Sky Survey.

In producing his *Atlas of Compact Groups of Galaxies*, the Canadian astronomer Paul Hickson focused on the set of small, relatively isolated, systems of galaxies with projected separations comparable to the diameters of the galaxies themselves.[5] Here is how the author explains the purpose of his work. "Consideration of the short apparent dynamical times of compact groups, as well as lingering questions concerning the frequency of galaxies with discordant redshifts in known groups prompted the author, in 1981, to begin a systematic program of study of these objects."

The atlas assembled a homogeneous sample of 100 compact groups of galaxies from a systematic search of the Palomar Sky Survey prints. Images of the groups were obtained at the Canada–France–Hawaii Telescope in blue and red band-passes, employing a CCD camera built by the author. The atlas included a catalogue, compiling information on photometric and spectroscopic parameters, along with other optical, infrared and radio data. Because several of Hickson's groups display signs of interaction, they had already been included in the collection of interacting galaxies by Arp and Vorontsov-Velyaminov.[6] Some images included objects not physically associated with the group, most likely due to chance alignments, as indicated by their discordant redshifts.

---

[4] A. H. Rots, A Neutral Hydrogen Mapping Survey of Large Galaxies, I – Observations, *Astronomy & Astrophysics Supplement Series*, 1980, Vol. 41, pp. 189–209.

[5] P. Hickson, *Atlas of Compact Groups of Galaxies*, Amsterdam: Gordon and Breach Science Publishers, 1994. It was also published in *Astro. Letters and Communications*, 1993, Vol. 29, pp. 1–207.

[6] B. Vorontsov-Velyaminov, *Atlas and Catalog of Interacting Galaxies*, Moscow: The Sternberg Astronomical Institute, Moscow University, 1959.

Surveying Our Neighbors

Another specialized atlas of galaxies is *An Atlas of Local Group Galaxies* by Paul Hodge, Brooke Skelton and Joy Ashizawa.[7] The focus of the atlas is the three dozen galaxies of the nearby group of galaxies to which our Milky Way belongs; it is called the Local Group. The atlas shows and describes 32 members of the Local Group; it excludes our Milky Way Galaxy, Messier 31, its companions and the nearby Magellanic Clouds (for which separate atlases had been published by the main author). Because the Local Group of galaxies are close to us, they can be imaged with a fine spatial scale, providing an enormous amount of detail, including resolved individual luminous stars and variable stars. It is indeed a remarkable feature of this atlas to be able to identify a large number of objects, such as emission regions (HII regions) and star clusters in each galaxy. The purpose of the atlas is stated on the back cover: "It is unique in its coverage and format and provides a source of these fundamental data that will be used for many years." Most unfortunately, the atlas suffered from poor reproduction by the publisher, which affected the quality of the work and most likely diminished its use, interest and impact.

Displaying Galaxy Fireworks

The Hubble Space Telescope has provided a number of spectacular images of galaxies in interaction or in the final process of merging. Lars Christensen, Davide de Martin and Raquel Shida created a magnificent teaching and pedagogical work, *Cosmic Collisions: The Hubble Atlas of Merging Galaxies*.[8] It is based on optical images with the fine spatial resolution dreamt of by Sandage and Bedke when they put together their ambitious *Atlas of Galaxies Useful for Measuring the Cosmological Distance Scale* in 1988. The images obtained by the many researchers using the Hubble Space Telescope are accessible through the Mikulski Archive of Space Telescopes (MAST) at the Space Telescope Science Institute. The MAST archive has been named in honor of the United States Senator Barbara Ann Mikulski from Maryland, who has been a staunch supporter of space sciences, science education and literacy.

The highly perturbed objects or weird galaxies illustrate vividly the cosmic collisions between galaxies, either merging or in highly disturbed states from a recent close passage. Zwicky, Ambartsumian and Sérsic would have greatly relished seeing so many galaxies in their "excited" states displayed in such a spectacular way. The authors have also written a fine introduction to the world of galaxies, including a general description of the Hubble scheme, stating that merging galaxies hold the clues for the history and future of galaxies.

Imaging Dust

The Spitzer Space Observatory provided researchers with an exceptional instrument to study galaxy morphology in the mid-infrared. It provided a tool to observe galaxies at longer

---

[7] P. W. Hodge, B. P. Skelton and J. Ashizawa, *An Atlas of Local Group Galaxies*, Dordrecht: Springer Netherlands, 2002.

[8] L. L. Christensen, D. de Martin and R. Y. Shida, *Cosmic Collisions: The Hubble Atlas of Merging Galaxies*, New York: Springer, 2009.

wavelengths where, with the appropriate use of filters, it becomes possible to disentangle stellar populations and the dust component very clearly. There have been several morphological studies in the infrared, but most have been selective and none very extensive. The sampling gap has now been filled by Ronald Buta and his colleagues, who published a large study of galaxies based on Spitzer imaging.[9]

Their article includes a full review of previous classification works and is supported by an on-line atlas with images of 207 galaxies at 3.6 microns.[10] The authors have chosen this wavelength because, seeing through the dust, it provides reddening-free sensitivity, mainly to older stars. Several key features of the de Vaucouleurs classification are particularly distinctive when galaxies are viewed at infrared wavelengths. The article explores the imprint of the de Vaucouleurs classification volume scheme. As the authors write, "3.6 μm classifications are well correlated with blue-light classifications, to the point where the essential features of many galaxies look very similar in the two very different wavelength regimes. Drastic differences are found only for the most dusty galaxies." A great advantage of the Spitzer 3.5-micron imaging is that it reaches deeper in brightness than other near-infrared imaging from the ground. This atlas, and the corresponding on-line material, gives the flavour of what the digital age adds to the power of scientific atlases.

A fine display of Spitzer infrared images of galaxies can be found on-line with the Spitzer Infrared Nearby Galaxies Survey (SINGS), which matches the Hubble tuning-fork classification.

An Atlas for Training

In order to fill a training need, the Australian astronomer Glen Mackie produced *The Multiwavelength Atlas of Galaxies*.[11] Its purpose is to depict how the best observed galaxies appear at different wavelengths, from gamma rays, X-rays, ultraviolet, optical, infrared to long radio wavelengths. It is clearly a pedagogical atlas for students and young researchers, showing well-known archetypal galaxies. What the small atlas lacks in homogeneity is compensated for in the display of diversity. The image gallery is supplemented with insights on the complex physical processes at play during various epochs of galaxy evolution.

Galaxies are illustrated over a range of dynamic states: normal, interacting, merging, starburst and active galaxies. The number of galaxies shown is small (34), but the objects have been observed over wavelengths spanning the whole electromagnetic spectrum. The atlas is a striking demonstration of the varying morphologies and shapes of galaxies between wavelengths, e.g. Cygnus A. In other cases, they differ less dramatically from one wavelength to another, e.g. Messier 101.

---

[9] R. Buta, et al., Mid-Infrared Galaxy Morphology from the Spitzer Survey of Stellar Structure in Galaxies (S$_4$G): The Imprint of the de Vaucouleurs Revised Hubble–Sandage Classification System at 3.6 micron, *Astrophysical Journal Supplement Series*, 2010, Vol. 190, pp. 147–165.

[10] http://kudzu.astr.ua.edu/s4gatlas/images.html

[11] G. Mackie, *The Multiwavelength Atlas of Galaxies*, Cambridge: Cambridge University Press, 2011.

Bibliography

Abt, Helmut A., *The Astrophysical Journal: American Astronomical Society Centennial Issue*, Chicago: University of Chicago Press & American Astronomical Society, 1999, Vol. 525, Number 1C, Part 3.

Aitken, R. G., Dorothea Klumpke Roberts: An Appreciation, *Publications of the Astronomical Society of the Pacific*, 1942, Vol. 54, p. 219.

Alexander, S., On the Origin of the Forms and the Present Condition of Some of the Clusters of Stars, and Several of the Nebulae. *The Astronomical Journal*, 1852, Vol. 2, pp. 95–160.

Alfvén, H. and Herlofson, N., Cosmic Radiation and Radio Stars, *Physical Review*, 1950, Vol. 78, p. 616.

Allington-Smith, J. R., Book Review: Atlas of Compact Groups of Galaxies, *The Observatory*, 1995, Vol. 115, p. 347.

Andrew, A. D., Book Review: Atlas of Compact Groups of Galaxies, *Irish Astronomical Journal*, 1995, Vol. 22, p. 226.

Arp, H. C., Atlas of Peculiar Galaxies, *The Astrophysical Journal Supplement Series*, 1966, Vol. 14, pp. 1–20, 57 pages of photographic plates.

Arp, H. C., *Atlas of Peculiar Galaxies*, Pasadena: California Institute of Technology, 1966.

Arp, H. C., Book Review: The Structure and Evolution of Galaxies, *Science*, 1966, Vol. 154, pp. 1439.

Arp, H. C. and Madore, Barry F., *A Catalogue of Southern Peculiar Galaxies and Associations*, Vols. 1 & 2, Cambridge: Cambridge University Press, 1987.

Baade, W., The Resolution of Messier 32, NGC 205, and the Central Region of the Andromeda Nebula, *The Astrophysical Journal*, 1944a, Vol. 100, pp. 137–146.

Baade, W., NGC 147 and NGC 185, Two New Members of the Local Group of Galaxies, *The Astrophysical Journal*, 1944b, Vol. 100, p. 148.

Baade, W., *Evolution of Stars and Galaxies*, ed. C. Payne-Gaposchkin, Cambridge, MA: Harvard University Press, 1963.

Baade, W. and Zwicky, F., On Super-Novae, *Proceedings of the National Academy of Sciences of the United States*, 1934, Vol. 20, pp. 254–259.

Babcock, H. W., The Rotation of the Andromeda Nebula, *Lick Observatory Bulletin*, 1939, Vol. XIX, No. 498, pp. 41–51.

Babcock, H. W., The Possibility of Compensating Astronomical Seeing, *Publications of the Astronomical Society of the Pacific*, 1953, Vol. 65, pp. 229–236.

Bacon, F., Aphorism 45, 55 and 109, *Novum Organum*, 1620.

Barnard, E. E., On the Dark Markings of the Sky with a Catalogue of 182 Such Objects, *The Astrophysical Journal*, 1919, Vol. 49, pp. 1–23

Barnard, E. E., *A Photographic Atlas of Selected Regions of the Milky Way*, ed. E. B. Frost and M. R. Calvert, Washington: Carnegie Institution of Washington, 1927.

Bartusiak, M., *The Day We Found the Universe*, New York: Vintage Books, 2010.

Bean, A., *Painting Apollo – First Artist on Another World*, Washington: Smithsonian Books, 2007.

Benson, M., Cosmigraphics, *Picturing the Universe Through Time*, New York: Abrams, 2014.

Berendzen, R., Hart, R. and Seely, D., *Man Discovers the Galaxies*, New York: Science History Publications, 1976.

Bettex, A., *The Discovery of Nature*, New York: Simon and Schuster, 1965.

Biagoli, M., *Galileo, Courtier: The Practice of Science in the Culture of Absolutism*, Chicago: University of Chicago Press, 1993.

Binney, J. and Merrifield, M., *Galactic Astronomy*, Princeton: Princeton University Press, 1998.

Bleichmar, D., The Geography of Observation: Distance and Visibility in Eighteenth-Century Botanical Travel, in *Histories of Scientific Observation*, ed. L. Daston and E. Lunbeck, Chicago: University of Chicago Press, 2011, pp. 373–395.

Block, D. L. and Freeman, K. C., *Shrouds of the Night, Masks of the Milky Way and Our Awesome New View of Galaxies*, New York: Springer, 2008.

Bond, G. P., An Account of the Nebula in Andromeda, *Memoirs of the American Academy of Arts and Sciences, New Series*, 1849, Vol. 3, pp. 75–86.

Bond, G. P., Stellar Photography, *Astronomische Nachrichten*, 1858, Vol. 49, pp. 81–100.

Bond, G. P., On the Spiral Structure of the Great Nebula in Orion, *Monthly Notices of the Royal Astronomical Society*, 1861, Vol. 21, pp. 203–207.

Bond, W. C., Bond, G. P. and Winlock, J., Results of Observations, *Annals of the Astronomical Observatory of Harvard College*, 1876, Vol. III.

Bredekamp, H., *Galilei Der Künstler: Der Mond, Die Sonne, Die Hand*, Berlin: Walter De Gruyter Inc., 2009.

Breidbach, O., *Visions of Nature: The Art and Science of Ernst Haeckel*, Munich: Prestel Publications, 2006.

Browers, L., Animal Vision Evolved 700 Million Years Ago, *Scientific American Blogs*, November 20, 2012.

Brush, S. G., *Nebulous Earth: The Origin of the Solar System and the Core of the Earth from Laplace to Jeffreys*, Cambridge: Cambridge University Press, 1996 (Three Volumes).

Burbidge, E. M., *Gérard de Vaucouleurs 1918–1995: A Biographical Memoir*, Washington: The National Academy Press, 2002, Vol. 82, pp. 1–17.

Buta, R., Galaxy Morphology and Classification, in *The World of Galaxies*, ed. H. G. Corwin Jr. and L. Bottinelli, New York: Springer-Verlag, 1989, pp. 29–44.

Buta, R. J., Galaxy Morphology, in *Planets, Stars, and Stellar Systems, Volume 6*, ed. W. C. Keel, New York: Springer, 2013, pp. 1–89.

Buta, R. J., Galaxy Morphology, in *Secular Evolution of Galaxies*, ed. Jesús Falcón-Barroso and Johann H. Knapen, Cambridge: Cambridge University Press, 2013, pp. 155–258.

Buta, R. J., Corwin Jr., H. G. and Odewahn, S. C., *The de Vaucouleurs Atlas of Galaxies*, Cambridge: Cambridge University Press, 2007.

Buta, R. J., Sheth, K., Regan, M., et al., Mid-Infrared Galaxy Morphology from the Spitzer Survey of Stellar Structure in Galaxies (S_4G): The Imprint of the de Vaucouleurs Revised Hubble–Sandage Classification System at 3.6 micron, *The Astrophysical Journal Supplement Series*, 2010, Vol. 190, pp. 147–165.

Capaccioli, M. and Corwin Jr., H. (eds.), *Gérard and Antoinette de Vaucouleurs: A Life for Astronomy*, Singapore: World Scientific, 1989.

Carrington, R., Description of a Singular Appearance seen in the Sun on September 1, 1859, *Monthly Notices of the Royal Astronomical Society*, 1859, Vol. 20, pp. 13–15.

Chapman, A. *The Victorian Amateur Astronomy: Independent Astronomical Research in Britain*, Hoboken: Wiley, 1999.

Chapman, A., William Parsons and the Irish Nineteenth-Century Tradition of Independent Astronomical Research, in *William Parsons, 3rd Earl of Rosse: Astronomy and the Castle in Nineteenth-Century Ireland*, ed. C. Mollan, 2014, Manchester: Manchester University Press, pp. 271–297.

Charmentier, I., Carl Linnaeus and the Visual Representation of Nature, *Historical Studies in the Natural Sciences*, 2011, Vol. 41, No. 4, 365–404.

Christensen, L. L., de Martin, D. and Shida, R. Y., *Cosmic Collisions: The Hubble Atlas of Merging Galaxies*, New York: Springer, 2009.

Common, A. A., Telescopes for Astronomical Photography, *Nature*, 1884, Vol. 31, pp. 38–40.

Corwin, H. G. Jr., Galaxy Catalogues and Surveys, in *The World of Galaxies*, ed. H. G. Corwin Jr. and L. Bottinelli, New York: Springer-Verlag, 1989, pp. 1–15.

Curtis, H. D., Three Novae in Spiral Nebulae, *Lick Observatory Bulletin*, Number 300, 1917, pp. 108–110.

Curtis, H. D., Descriptions of 762 Nebulae and Clusters Photographed with the Crossley Reflector, *Publications of the Lick Observatory*, 1918, Vol. XIII, Part I, pp. 9–42.

Curtis, H. D., A Study of Occulting Effects in the Spirals, *Publications of the Lick Observatory*, 1918, Vol. XIII, Part II, pp. 43–55.

Curtis, H. D., The Number of the Spiral Nebulae, *Publications of the Astronomical Society of the Pacific*, 1918, Vol. 30, pp. 159–161.

Curtis, H. D., Modern Theories of the Spiral Nebulae, *Journal of the Washington Academy of Sciences*, 1920, Vol. 9, p. 217.

Danjon, A. and Couder, A., *Lunettes et télescopes*, Paris: Librairie scientifique et technique Albert Blanchard, 1999, p. 694 (original work published in 1935).

Daston, L., On Scientific Observations, *Isis*, 2008, Vol. 99, No. 1, pp. 97–110.

Daston, L. and Galison, P., The Image of Objectivity, *Representations*, No. 40, 1992, pp. 81–128.

Daston, L. and Galison, P., *Objectivity*, New York: Zone Books, 2007.

Daston, L. and Lunbeck, E. (eds.), *Histories of Scientific Observation*, Chicago: University of Chicago Press, 2011.

Daston, L. and Lunbeck, E., Observing Together: Communities, in *Histories of Scientific Observation*, Chicago: University of Chicago Press, 2011, pp. 369–371.

Dawkins, R., *The Blind Watchmaker*, New York: W. W. Norton & Company, 1996.

Dejaiffe, R., Book Review, *Ciel et Terre*, 2007, Vol. 123, p. 121.

de Vaucouleurs, G., Survey of Bright Galaxies South of −35° Declination, with the 30-Inch Reynolds Reflector (1952–1955), *Memoirs of the Commonwealth Observatory*, 1956, Vol. III, No. 13, pp. 1–84.

de Vaucouleurs, G., Classifying Galaxies, *Astronomical Society of the Pacific Leaflets*, No. 341, 1957, pp. 329–336.

de Vaucouleurs, G., Classification, Dimensions, and Distances of Bright Southern Galaxies, *Sky and Telescope*, 1957, Vol. 16, p. 582.

de Vaucouleurs, G., Classification and Morphology of External Galaxies, *Handbuch der Physik*, 1959, Vol. 53, pp. 275–310.

de Vaucouleurs, G., Algunos comentarios sobre fotometria de galaxias y el "Atlas de galaxies australes" de Jose Lui Sérsic, *Revista Astronomica*, 1969, Vol. XLI, pp. 27–34.

de Vaucouleurs, G., de Vaucouleurs, A., Corwin Jr., H. G., Buta, R. J., Paturel, G. and Fouque, P., *Third Reference Catalogue of Bright Galaxies*, New York: Springer-Verlag, 1991.

Dewhirst, D. W. and Hoskin, M., The Rosse Spirals, *Journal for the History of Astronomy*, 1991, Vol. XXII, pp. 257–266.

Dick, S. J., *Discovery and Classification in Astronomy, Controversy and Consensus*, Cambridge: Cambridge University Press, 2013.

Dreyer, J. L. E. (ed.), *The Scientific Papers of Sir William Herschel*, London: Royal Astronomical Society, 1912 (Two Volumes).

Easton, C., A Photographic Chart of the Milky Way and the Spiral Theory of the Galactic System, *The Astrophysical Journal*, 1913, Vol. 37, p. 105.

Eddington, A. S., On the Relation Between the Masses and Luminosities of the Stars, *Monthly Notices of the Royal Astronomical Society*, 1924, Vol. 84, pp. 308–332.

Edmondson, F., Reconnaissance of Outer Space, *Science*, 1961, Vol. 134, p. 464.

Elder, F. R., Gurewitsch, A. M., Langmuir, R. V. and Pollock, H. C., Radiation from Electrons in a Synchrotron, *Physical Review*, 1947, Vol. 71, pp. 829–830.

Elmegreen, D. M. and Elmegreen, B. G., Arm Classifications for Spiral Galaxies, *The Astrophysical Journal*, 1987, Vol. 313, pp. 3–9.

Elphick, J., *Birds, The Art of Ornithology*, New York: Rizzoli International Publications, Inc., 2015.

Ewen, H. I. and Purcell, E. M., Radiation from Galactic Hydrogen at 1,420 MHz, *Nature*, 1951, Vol. 168, p. 356.

Fabricant, D., Lecar, M. and Gorenstein, P., X-ray Measurements of the Mass of M87, *The Astrophysical Journal*, 1980, Vol. 241, pp. 552–560.

Fath, E. A., The Spectra of Some Spiral Nebulae and Globular Star Clusters, *Lick Observatory Bulletin*, 1909, No. 149, pp. 71–77.

Fermi, E., Galactic Magnetic Fields and the Origin of Cosmic Radiation, *The Astrophysical Journal*, 1954, Vol. 119, pp. 1–6.

Fernie, J. D., The Historical Quest for the Nature of the Spiral Nebulae, *Publications of the Astronomical Society of the Pacific*, 1970, Vol. 82, pp. 1189–1230.

Ferris, T., *Galaxies*, New York: Random House Publishing, 1988.

Field, G., Book Review: The Hubble Atlas of Galaxies, *American Scientist*, 1962, Vol. 50, p. 212A.

Fisher, C., *The Golden Age of Flowers: Botanical Illustration in the Age of Discovery 1600–1800*, London: The British Library, 2011.

Fox, W. L., *Alan Bean Painting Apollo*, Washington: Smithsonian Books, 2007.

Fourier, J.-B., *Théorie analytique de la chaleur*, Paris: Chez Firmin Didot, père et fils, 1822.

Freeman, K. C., On the Disks of Spiral and S0 Galaxies, *The Astrophysical Journal*, 1970, Vol. 160, pp. 811–830.

Freese, K., *The Cosmic Cocktail: Three Parts of Dark Matter*, Princeton: Princeton University Press, 2014.

Galilei, G., *Istoria e Demostrazioni Intorno alle Macchie Solaris e Loro Accidenti*, Rome: Giacomo Mascardi, 1613.

Galison, P., *Image and Logic: A Material Culture of Microphysics*, Chicago: University of Chicago Press, 1997.

Geller, M., Book Review, *Science*, 1995, Vol. 268, p. 1214.

Gendler, R., Forays into Astronomical Imaging: One Person's Experience and Perspective, *Astronomy Beat, Astronomical Society of the Pacific*, No. 79, August 30, 2011, pp. 1–6.

Giacconi, R., *Secrets of the Hoary Deep, A Personal History of Modern Astronomy*, Baltimore: Johns Hopkins University Press, 1998.

Gilmore, G., Book Review, *The Observatory*, 1995, Vol. 115, No. 1128, p. 278.

Gingerich, O. Through Rugged Ways to the Galaxies, *Journal for the History of Astronomy*, 1990, Vol. 21, p. 79.

Glass, I. S., *Victorian Telescope Makers, The Lives and Letters of Thomas and Howard Grubb*, Bristol: Institute of Physics Publishing, 1997.

Goodstein, J. R., *Millikan's School*, New York: W. W. Norton & Company, 1991.

Gordin, M. D., *A Well-Ordered Thing, Dmitrii Mendeleev and the Shadow of the Periodic Table*, New York: Basic Books, 2004, p. xvii.

Gould, J., *A Monograph of the Trochilidae, or Family of Humming-Birds*, London: Henry Sotheran & Co., 1880.

Gould, S. J., Why Darwin?, *The New York Review of Books*, April 14, 1996, Vol. XLIII, No. 6, p. 10.

Harwit, M., *In Search of the True Universe: The Tools, Shaping and Cost of Cosmological Thought*, Cambridge: Cambridge University Press, 2013.

Havens, E. (ed.), *The Dr. Elliott & Eileen Hinkes Collection of Rare Books in the History of Scientific Discovery*, Baltimore: The Sheridan Libraries, Johns Hopkins University, 2011.

Hearnshaw, J. B., *The Analysis of Starlight: One Hundred and Fifty Years of Astronomical Spectroscopy*, Cambridge: Cambridge University Press, 1986.

Herschel, J. F. W., Observations of Nebulae and Clusters of Stars, Made at Slough, with a Twenty-Feet Reflector, between the Years 1825 and 1833, *Philosophical Transactions of the Royal Society*, 1833, Vol. 123, pp. 359–505.

Herschel, J. F. W., Report of the Fifteenth Meeting of the British Association for the Advancement of Science Held at Cambridge in June 1845, London, 1846, p. xxxvi.

Herschel, J. F. W., *Results of Astronomical Observations Made During the Years 1834, 5, 6, 7, 8 at the Cape of Good Hope*, London: Smith, Elder & Co., 1847.

Herschel, J. F. W., General Catalogue of Nebulae and Clusters of Stars, *Philosophical Transactions of the Royal Society*, 1864, Vol. 154, pp. 1–137.

Herschel, W., On the Construction of Heavens, *Philosophical Transactions of the Royal Society*, Vol. 75, 1785, pp. 213–266.

Herschel, W., Astronomical Observations relating to the Construction of the Heavens, *Philosophical Transactions of the Royal Society*, 1811, Vol. 101, pp. 269–336.

Hey, J. S., Solar Radiation in the 4–6 Metre Radio Wave-Length Band, *Nature*, 1946, Vol. 157, pp. 47–48.

Hickson, P., *Atlas of Compact Groups of Galaxies*, Amsterdam: Gordon and Breach Science Publishers, 1994.

Hirshfeld, A., *Parallax: The Race to Measure the Cosmos*, New York: W. H. Freeman and Company, 2001.

Hirshfeld, A., *Starlight Detectives: How Astronomers, Inventors and Eccentrics Discovered the Modern Universe*, New York: Bellevue Literary Press, 2014.

Hockey, T., Trimble, V. and Williams, T. R. (eds.), *Biographical Encyclopedia of Astronomers*, New York: Springer, 2007.

Hodge, P. W., Skelton, B. P. and Ashizawa, J., *An Atlas of Local Group Galaxies*, Dordrecht: Springer, 2002.

Holmberg, E., A Photometric Study of Nearby Galaxies, *Meddelanden fran Lunds Astronomiska Observatorium*, 1950, Ser. II, 128, pp. 1–13.

Hoskin, M., The "Great Debate": What Really Happened, *Journal for the History of Astronomy*, 1976, Vol. 7, pp. 169–182.

Hoskin, M., The Cosmology of Thomas Wright of Durham, in *Stellar Astronomy: Historical Studies*, Chalfont St. Giles: Science History Publications, 1982, pp. 113–114.

Hoskin, M., The First Drawing of a Spiral Nebula, *Journal for the History of Astronomy*, 1982, Vol. XIII, pp. 97–101.

Hoskin, M., Caroline Herschel as Observer, *Journal for the History of Astronomy*, Vol. XXXVI, 2005, p. 394.

Hoskin, M., *Discoverers of the Universe: William and Caroline Herschel*, Princeton: Princeton University Press, 2011.

Hoskin, M., *The Construction of the Heavens: William Herschel's Cosmology*, Cambridge: Cambridge University Press, 2012.

Houston, W. S., Book Review: Atlas of Galaxies Useful for Measuring the Cosmological Distance Scale, *Sky and Telescope*, 1989, Vol. 78, p. 40.

Hubble, E. P., Extra-Galactic Nebulae, *Contributions from the Mount Wilson Observatory*, Washington: Carnegie Institution of Washington, 1926, Vol. 324, pp. 1–49.

Hubble, E. P., Extra-Galactic Nebulae, *The Astrophysical Journal*, 1926, Vol. 64, pp. 321–369.

Hubble, E. P., A Relation Between Distance and Radial Velocity among Extra-Galactic Nebulae, *Proceedings of the National Academy of Sciences of the United States*, 1929, Vol. 15, pp. 168–173.

Hubble, E. P., The Distribution of Extragalactic Nebulae, *The Astrophysical Journal*, 1934, Vol. 79, pp. 8–76.

Hubble, E. P., *The Realm of the Nebulae*, New Haven: Yale University Press, 1936.

Hubble, E. P. and Humason, M. L., The Velocity-Distance Relation Among Extra-Galactic Nebulae, *The Astrophysical Journal*, 1931, Vol. 74, pp. 43–80.

Huggins, W. and Huggins, Lady M., *The Scientific Papers of Sir William Huggins*, London: Wesley & Son, 1909.

Humason, M., The Apparent Velocities of 100 Extra-Galactic Nebulae, *The Astrophysical Journal*, 1936, Vol. 83, pp. 10–22.

Jaki, S. L., *The Milky Way, An Elusive Road for Science*, New York: Science History Publications, 1972.

Jansky, K. G., Electrical Disturbances Apparently of Extraterrestrial Origin, *Proceedings of the Institute of Radio Engineers*, 1933, Vol. 21, p. 1387.

Jansky, K. G., Note on the Source of Interstellar Interference, *Proceedings of the Institute of Radio Engineers*, 1935, Vol. 23, pp. 1158–1163.

Jeans, J., *Astronomy and Cosmogony*, Cambridge: Cambridge University Press, 1928.

Johnson, G., *Miss Leavitt's Stars: The Untold Story of the Woman Who Discovered How to Measure the Universe*, New York: W. W. Norton & Company, 2005.

Kanipe, J. and Webb, D., *The Arp Atlas of Peculiar Galaxies: A Chronicle and Observer's Guide*, Richmond: Willmann-Bell, Inc., 2007.

Kant, I., *Universal Natural History and Theory of the Heavens*, Transl. I. Johnston, Arlington: Richer Resources Publications, 2008.

Keeler, J., Note on a Case of Differences Between Drawings and Photographs of Nebulae, *Publications of the Astronomical Society of the Pacific*, 1895, Vol. 7, pp. 279–282.

Keeler, J., The Crossley Reflector of the Lick Observatory, *Publications of the Astronomical Society of the Pacific*, 1900, Vol. 12, pp. 146–167.

Kellerman, K. I., Grote Reber's Observations of Cosmic Static, *The Astrophysical Journal: American Astronomical Society Centennial Issue*, 1999, Vol. 525, p. 372.

King, H. C., *The History of the Telescope*, Mineola: Dover Publications, Inc., 2003.

Kodaira, K., Okamura, S. and Ichikawa, S.-I., *Photometric Atlas of Northern Bright Galaxies*, Tokyo: University of Tokyo Press, 1990.

Kormendy, J., Observations of Galaxy Structure and Dynamics, in *Morphology and Dynamics of Galaxies: Proceedings of the 12th Advanced Course*, Saas-Fee: Observatoire de Genève, 1983, pp. 113–288.

Kormendy, J. and Bender, R., A Revised Parallel-Sequence Morphological Classification of Galaxies: Structure and Formation of S0 and Spheroidal Galaxies, *The Astrophysical Journal Supplement Series*, 2012, Vol. 198, pp. 1–40.

Kragh, H. S., *Conceptions of Cosmos, From Myths to the Accelerating Universe: A History of Cosmology*, Oxford: Oxford University Press, 2007.

Kunitzsch, P., Medieval Reference to the Andromeda Nebula, *The Messenger*, 1987, vol. 49, pp. 42–43.

Kunitzsch, P., Sufi: Abu al-Husayn Abd al-Rahman ibn Umar al-Sufi, in *Biographical Encyclopedia of Astronomers*, eds. T. Hockey, V. Trimble and T. R. Williams, New York: Springer, 2007, p. 1110.

Lang, K. R. and Gingerich, O. (eds.), *A Source Book in Astronomy and Astrophysics, 1900–1975*, Cambridge, MA: Harvard University Press, 1979.

Laplace, P. S., *Exposition du système du monde*, Paris: Courcier, Quai des Augustines, 1808.

Laplace, P. S., *The System of the World*, Vols. 1 & 2, London: Printed for Richard Phillips, 1809.

Lasker, B. M., Lattanzi, M. G., McLean, B. J., et al., The Second-Generation Guide Star Catalog: Description and Properties, *The Astronomical Journal*, 2008, Vol. 136, pp. 735–776.

Leavitt, H. S., 1777 Variables in the Magellanic Clouds, *Annals of the Harvard College Observatory*, 1908, Vol. 60, pp. 87–108.

Leavitt, H. S. and Pickering, E. C., Periods of Twenty-Five Variable Stars in the Small Magellanic Cloud, *Harvard College Observatory Circulars*, No. 173, 1912, pp. 1–3.

Lemaître, G., Un univers homogène de masse constante et de rayon croissant, rendant compte de la vitesse radiale des nébuleuses extra-galactiques, *Annales de la Société Scientifique de Bruxelles*, 1927, Vol. 47, pp. 49–59.

Lemaître, G., The Beginning of the World from the Point of View of Quantum Theory, *Nature*, 1931, Vol. 127, p. 706.

Lequeux, J., *François Arago: un savant généreux – Physique et astronomie au XIX siècle*, Paris: EDP Sciences/L'Observatoire de Paris, 2008.

Linnaeus, C., *Species Plantarum*, Stockholm: Holmiae, Impensis Laurentii Salvii, 1753.

Livio, M., Mystery of the Missing Text Solved, *Nature*, 2011, Vol. 479. pp. 171–173.

Lovell, B., *The Story of Jodrell Bank*, New York: Harper & Row, 1968.

Lyell, C., *Principles of Geology*, London: Penguin Books, 1997.

Lynch, M. and Edgerton Jr., S. Y., Aesthetic and Digital Image Processing: Representational Craft in Contemporary Astronomy, *The Sociological Review*, 1987, Vol. 35, pp. 184–220.

Lynden-Bell, D. and Schweizer, F., Allan Rex Sandage, 18 June 1926–13 November 2010, *Biographical Memoirs of Fellows of the Royal Society*, 2012 (arXiv:1111.5646).

Lynds, B. T., An Atlas of Dust and HII Regions in Galaxies, *The Astrophysical Journal Supplement Series*, 1974, Vol. 28, pp. 391–406.

Mackie, G., *The Multiwavelength Atlas of Galaxies*, Cambridge: Cambridge University Press, 2011.

Malin, D. and Murdin, P., *Colours of the Stars*, Cambridge: Cambridge University Press, 1984.

Malin, D. F., Astronomical Photography, *Gérard and Antoinette de Vaucouleurs: A Life for Astronomy*, Singapore: World Scientific, 1989, pp. 53–64.

Malmquist, G., On Some Relations in Stellar Statistics, *Arkiv för Mathematik, Astronomi och Fysik*, 1922, Vol. 16, pp. 1–52.

Mathewson, D. S. and Ford, V. L., Polarization Observations of 1800 stars, *Memoirs of the Royal Astronomical Society*, 1970, Vol. 74, pp. 139–182.

Mihalas, D., Baade's Resolution of M32, NGC 205, and M31, in *The Astrophysical Journal: American Astronomical Society Centennial Issue*, 1999, Vol. 525, Number IC, Part 3, pp. 359–360.

Mollan, C., A Consummate Engineer, in *William Parsons, 3rd Earl of Rosse: Astronomy and the Castle in Nineteenth-Century Ireland*, ed. C. Mollan, Manchester: Manchester University Press, 2014, pp. 159–209.

Mollan, C. (ed.), *William Parsons, 3rd Earl of Rosse: Astronomy and the Castle in Nineteenth-Century Ireland*, Manchester: Manchester University Press, 2016.

Moore, S. L., Book Review: The de Vaucouleurs Atlas of Galaxies, *Journal of the British Astronomical Association*, 2007, Vol. 117, no. 4, p. 211.

Morgan, W. W., Some Characteristics of Galaxies, *The Astrophysical Journal*, 1962, Vol. 135, pp. 1–10.

Morgan, W. W., A Morphological Life, *Annual Review of Astronomy and Astrophysics*, 1988, Vol. 26, pp. 1–10.

Muller, C. Alex and Oort, Jan H., The Interstellar Hydrogen Line at 1,420 MHz and an Estimate of Galactic Rotation, *Nature*, 1951, Vol. 168, pp. 356–358.

Nasim, O. W., *Observing by Hand: Sketching Nebulae in the Nineteenth Century*, Chicago: University of Chicago Press, 2013.

Nicolaidis, E., *Science and Eastern Orthodoxy: From the Greek Fathers to the Age of Globalization*, Baltimore: Johns Hopkins University Press, 2011.

North, J., *Cosmos: An Illustrated History of Astronomy and Cosmology*, Chicago: University of Chicago Press, 2008.

Nussbaumer, H. and Bieri, L., *Discovering the Expanding Universe*, Cambridge: Cambridge University Press, 2010.

Oort, J. H., Observational Evidence confirming Lindblad's Hypothesis of a Rotation of the Galactic System, *Bulletin of the Astronomical Institutes of the Netherlands*, 1927, Vol. 3, pp. 275–282.

Oort, J. H., The Force Exerted by the Stellar System in the Direction Perpendicular to the Galactic Plane and Some Related Problems, *Bulletin of the Astronomical Institutes of the Netherlands*, 1932, Vol. VI, pp. 249–287.

Oort, J. H., Some Problems Concerning the Structure and Dynamics of the Galactic System and the Elliptical Nebulae NGC 3113 and 4494, *The Astrophysical Journal*, 1940, Vol. 91, pp. 273–306.

Oort, J. H., Kerr, F. J. and Westerhout, G., The Galactic System as a Spiral Nebulae, *Monthly Notices of the Royal Astronomical Society*, 1958, Vol. 118, 379–389.

Osterbrock, D. E., *Pauper & Prince: Ritchey, Hale & Big American Telescopes*, Tucson: The University of Arizona Press, 1993.

Osterbrock, D. E., *Yerkes Observatory 1892–1950: The Birth, Near Death, and Resurrection of a Scientific Research Institution*, Chicago: University of Chicago Press, 1997.

Osterbrock, D. E., *Walter Baade, A Life in Astrophysics*, Princeton: Princeton University Press, 2011.

Parsons, C. (ed.), *The Scientific Papers of the Third Earl of Rosse 1800–1867*, Cambridge: Cambridge University Press, Original edition 1926, Digitally printed version 2010.

Peale, T., *The Butterflies of North America, Titian Peale's Lost Manuscript*, New York: American Museum of Natural History/Abrams, 2015.

Proctor, R. A., The Rosse Telescope Set to New Work, *Frazer's Magazine for Town and Country*, 1869, Vol. LXXX, pp. 754–760.

Ptolemy's Almagest, Transl. G. J. Toomer, Princeton: Princeton University Press, 1998.

Randall, L., *Dark Matter and the Dinosaurs: The Astounding Interconnectedness of the Universe*, New York: HarperCollins Publishers, 2015.

Reber, G., Cosmic Static, *The Astrophysical Journal*, 1944, Vol. 100, pp. 279–287.

Rector, T. A., Levay, Z., Frattare, L. M., English, J., and Pu'uohau-Pummill, K., Image-Processing Techniques for the Creation of Presentation-Quality Astronomical Images, *Astronomical Journal*, 2007, Vol. 133, pp. 598–611.

Reeves, E. and Van Helden, A., *On Sunspots: Galileo Galilei & Christoph Scheiner*, Transl. E. Reeves and A. van Helden, Chicago: University of Chicago Press, 2010.

Reinmuth, K., Die Herschel-Nebel nach Aufnahmen der Künigstuhl-Sternwarte, *Veroeffentlichungen der Badischen Sternwarte zu Heidelberg*, Berlin: De Gruyter, 1926, Vol. VI.

Reynolds, J. H., Photometric Measure of the Nuclei of Some Typical Spiral Nebulae, *Monthly Notices of the Royal Astronomical Society*, 1920, Vol. 80, pp. 746–753.

Ritchey, G., Novae in Spiral Nebulae, *Publications of the Astronomical Society of the Pacific*, 1917, Vol. 29, pp. 210–212.

Ritchey, G., The Modern Reflecting Telescope and the New Astronomical Photography, *Transactions of the Optical Society*, 1928, Vol. 29, p. 197.

Roberts, I., Photographs of the Nebulae M31, h44, and h51 Andromedae, and M27 Vulpeculae, *Monthly Notices of the Royal Astronomical Society*, 1888, Vol. 49, p. 65.

Roberts, I., Photographs of the Nebulae in the Pleiades and in Andromeda, *Monthly Notices of the Royal Astronomical Society*, 1889, Vol. 49, pp. 120–121.

Roberts, I., *A Selection of Photographs of Stars, Star-clusters and Nebulae, Volume I*, London: The Universal Press, 1893.

Roberts, I., Photograph of the Spiral Nebula M33 Trianguli, *Monthly Notices of the Royal Astronomical Society*, 1895, Vol. 56, 70–71.

Roberts, I., *A Selection of Photographs of Stars, Star-clusters and Nebulae, Volume II*, London: "Knowledge" Office, 1899.

Roberts, M. S. and Whitehurst, R. N., The Rotation Curve and Geometry of M31 at Large Galactocentric Distances, *The Astrophysical Journal*, 1975, Vol. 201, pp. 327–346.

Rosse, Third Earl of, On the Construction of Specula of Six-Feet Aperture: And a Selection from the Observations of Nebulae Made with Them, *Philosophical Transactions of the Royal Society*, 1861, Vol. 151, pp. 681–745.

Rosse, Fourth Earl of, An Account of the Observations on the Great Nebula in Orion, Made at Birr Castle, with the 3-feet and 6-feet Telescope, Between 1848 and 1867, With a Drawing of the Nebula, *Philosophical Transactions of the Royal Society*, 1868, Vol. 158, pp. 57–73.

Rosse, Fourth Earl of, Observations of Nebulae and Clusters of Stars Made with the Six-foot and Three-foot Reflectors at Birr Castle, From the Year 1848 Up to About the Year 1878, *The Scientific Transactions of the Royal Dublin Society*, 1879, Vol. 11.

Rosse, A., Countess of, Mary, Countess of Rosse (1813–85), in *William Parsons, 3rd Earl of Rosse: Astronomy and the Castle in Nineteenth-Century Ireland*, ed. C. Mollan, Manchester: Manchester University Press, 2014, pp. 55–70.

Rots, A., *Atlas of 21 cm HI line Profiles for 61 Galaxies of Large Angular Size*, Green Bank: National Radio Astronomy Observatory, 1979.

Rubin, V., Book Review: A Revised Shapley–Ames Catalog of Bright Galaxies, *Sky and Telescope*, 1982, Vol. 63, p. 478.

Rudaux, L. and de Vaucouleurs, G., *Larousse Encyclopedia of Astronomy*, New York: Prometheus Press, 1959.

Rudwick, M. J. S., The Emergence of a Visual Language for Geological Science 1740–1840, *History of Science*, 1976, Vol. XIV, pp. 149–195.

Rudwick, M. J. S., *The Great Devonian Controversy: The Shaping of Scientific Knowledge among Gentlemanly Specialists*, Chicago: University of Chicago Press, 1985.

Rudwick, M. J. S., *Bursting the Limits of Time: The Reconstruction of Geohistory in the Age of Revolution*, Chicago: University of Chicago Press, 2005.

Rudwick, M. J. S., *Earth's Deep History: How It Was Discovered and Why It Matters*, Chicago: University of Chicago Press, 2014.

Ryle, M., The New Cambridge Radio Telescope, *Nature*, 1962, Vol. 194, pp. 517–518.

Sandage, A. R., *The Hubble Atlas of Galaxies*, Washington: Carnegie Institution of Washington Publications, No. 618, 1961.

Sandage, A. R., Classification and Stellar Content of Galaxies Obtained from Direct Photography, in *Galaxies and the Universe*, eds. A. Sandage, M. Sandage and J. Kristian, Chicago: University of Chicago Press, 1975, pp. 1–55.

Sandage, A. R., Star formation rates, galaxy morphology, and the Hubble sequence, *Astronomy and Astrophysics*, 1986, Vol. 161, pp. 89–101.

Sandage, A. R., *Centennial History of the Carnegie Institution of Washington, Volume I: The Mount Wilson Observatory*, Cambridge: Cambridge University Press, 2004.

Sandage, A. R. and Tammann, G. A., *The Revised Shapley-Ames Catalog of Bright Galaxies*, Washington: Carnegie Institution of Washington Publications, 1981.

Sandage, A. R. and Bedke, J., *Atlas of Galaxies Useful for Measuring the Cosmological Distance Scale*, Washington: Scientific and Technical Information Division, NASA, 1988.

Sandage, A. R. and Bedke, J., *The Carnegie Atlas of Galaxies*, Washington: Carnegie Institution of Washington Publications, 1994, Volumes I and II.

Schaffer, Simon, On Astronomical Drawing, in *Picturing Science, Producing Art*, ed. C. A. Jones and P. Galison, New York: Routledge, 1998, pp. 441–474.

Scheiner, J., On the Spectrum of the Great Nebula in Andromeda, *The Astrophysical Journal*, 1899, Vol. 9, pp. 149–150.

Scheuer, P. A. G., The Development of Aperture Synthesis at Cambridge, in *The Early Years of Radio Astronomy*, ed. W. T. Sullivan III, Cambridge: Cambridge University Press, 1984, pp. 249–265.

Schmidt, M., A Model of the Distribution of Mass in the Galactic System, *Bulletin of the Astronomical Institutes of the Netherlands*, 1956, Vol. 13, pp. 211–222.

Schultz, D., *The Andromeda Galaxy and the Rise of Modern Astronomy*, New York: Springer, 2012.

Schweizer, F., Observational Evidence for Interactions and Mergers, in *Galaxies: Interactions and Induced Star Formation*, eds. R. C. Kennicutt Jr. et al., Saas-Fee Advanced Course 26, Swiss Society for Astrophysics and Astronomy, 1996, pp. 105–274.

Sérsic, J. L., *Atlas de galaxias australes*, Cordoba: Observatorio Astronómico, 1968.

Seyfert, C. K., Nuclear Emission in Spiral Nebulae, *The Astrophysical Journal*, 1943, Vol. 97, pp. 28–40.

Shapin, S. and Schaffer, S., *Leviathan and the Air Pump, Hobbes, Boyle, and the Experimental Life*, Princeton: Princeton University Press, 1985, 2011.

Shapley, H., Studies Based on the Colors and Magnitudes in Stellar Clusters – Seventh Paper: The Distances, Distribution in Space and Dimensions of 69 Globular Clusters, *The Astrophysical Journal*, 1918, Vol. 48, pp. 154–181.

Shapley, H. and Ames, A., A Study of a Cluster of Bright Spiral Nebulae, *Harvard College Observatory Circular*, 1926, Vol. 294, pp. 1–8.

Shapley, H. and Ames, A., A Survey of the External Galaxies Brighter Than the Thirteenth Magnitude, *Annals of the Astronomical Observatory of Harvard College*, 1932, Vol. 88, pp. 41–76.

Shapley, H. and Paraskevopoulos, J. S., Galactic and Extragalactic Studies, III: Photographs of Thirty Southern Nebulae and Clusters, *Proceedings of the National Academy of Sciences of the United States*, 1940, Vol. 26, pp. 31–36.

Sharpe, T., The birth of the geological map, *Science*, 2015, Vol. 347, pp. 230–232.

Shears, J., Knight, C., Lewis, M., Macdonald, L., Moore, S. and Young, J., In the Footsteps of Ebenezer Porter Mason and His Nebulae, 2014arXiv:1401.7960.

Shklovsky, I. S., *Cosmic Radio Waves*, Cambridge, MA: Harvard University Press, 1960.

Slipher, V. M., Nebulae, *Proceedings of the American Philosophical Society*, 1917, Vol. 56, pp. 403–409.

Smith, R., *The Expanding Universe, Astronomy's "Great Debate" 1900–1931*, Cambridge: Cambridge University Press, 1982, pp. 220–270.

Smothers, R., Commemorating a Discovery in Radio Astronomy, *New York Times*, June 9, 1998.

Snider, E., The Eye of Hubble, Framing Astronomical Images, *Frame: A Journal of Visual and Material Culture*, Issue One, Spring 2011, pp. 3–21.

Sobel, D., *The Glass Universe, How the Ladies of the Harvard Observatory Took the Measure of the Stars*, New York: Viking, 2016.

Steinicke, W., *Observing and Cataloguing Nebulae and Star Clusters: From Herschel to Dreyer's New General Catalogue*, Cambridge: Cambridge University Press, 2010.

Steinicke, W., Birr Castle Observations of Non-Stellar Objects and the Development of Nebular Theories, in *William Parsons, 3rd Earl of Rosse: Astronomy and the Castle in Nineteenth-Century Ireland*, ed. C. Mollan, Manchester: Manchester University Press, 2014, p. 210.

Stigler, S. M., Stigler's Law of Eponymy, *Transactions of the New York Academy of Sciences*, 1980, Vol. 38, pp. 147–158.

Stöckli, A. and Müller, R., *Fritz Zwicky: An Extraordinary Astrophysicist*, Cambridge: Cambridge Scientific Publishers, 2011.

Stone, R. P. S., The Crossley Reflector: A Centennial Review – I & II, *Sky and Telescope*, 1979, October issue, pp. 307–311; November issue, pp. 396–400.

Stoney, G. J., in 4th Earl of Rosse, On the Construction of Specula of Six-Feet Aperture: And a Selection from the Observations of Nebulae Made with Them, *Philosophical Transactions of the Royal Society*, 1861, Vol. 151, Appendix.

Størmer, C., *Photographic Atlas of Auroral Forms*, Oslo: International Geodetic and Geophysical Union, 1930.

Struve, O. and Zebergs, V., *Astronomy of the 20th Century*, New York: The Macmillan Company, 1962.

Sullivan, III, W. T., Some Highlights of Interferometry in Early Radio Astronomy, in *Radio Inteferometry: Theory, Techniques and Applications*, eds. T. J. Cornwell and R. A. Perley, 1991, Astronomical Society of the Pacific Conference Series, Vol. 19, pp. 132–149.

Takase, B., Kodaira, K. and Okamura, S., *An Atlas of Selected Galaxies: With Illustrations of Photometric Analyses*, Tokyo: University of Tokyo Press; Utrecht, the Netherlands: VNU Science Press, 1984.

Teerikorpi, P., Lundmark's Unpublished 1922 Nebula Classification, *Journal for the History of Astronomy*, Science History Publications, 1989, Vol. 20, pp. 165–170.

Tinsley, B. J., Evolution of the Stars and Gas in Galaxies, *The Astrophysical Journal*, 1968, Vol. 151, pp. 547–565.

Toomre, A. and Toomre, U., Galactic Bridges and Tails, *The Astrophysical Journal*, 1972, Vol. 178, pp. 623–666.

Tropp, E. A., Frenkel, V. Ya. and Chernin, A. D., *Alexander A. Friedmann: The Man who Made the Universe Expand*, Cambridge: Cambridge University Press, 1993.

Trouvelot, E. L., *Astronomical Sketches Taken at the Harvard College Observatory 1878–18*. Originals held at the Harvard University John Wolbach Library.

Trumpler, R. J., Preliminary Results on the Distances, Dimensions and Space Disribution of Open Star Clusters, *Lick Observatory Bulletin*, No. 42, 1930, pp. 154–188.

Trumpler, R. J., Absorption of Light in the Galactic System, *Publications of the Astronomical Society of the Pacific*, 1930, Vol. 42, pp. 214–227.

Tufte, E., *The Visual Display of Quantitative Information*, Cheshire: Graphics Press, 1992.

van de Hulst, H. C., Radio Waves from Space: Origin of Radio Waves, *Nederlands tijdschrift voor natuurkunde*, 1945, Vol. 11, pp. 210–221.

van den Bergh, S., A Preliminary Luminosity Classification of Late-Type Galaxies, and A Preliminary Luminosity Classification for Galaxies of Type Sb, *The Astrophysical Journal*, 1960, Vol. 131, pp. 215–223 and pp. 558–573.

van den Bergh, S., Review of Publications: The Hubble Atlas of Galaxies, *Journal of the Royal Astronomical Society of Canada*, 1962, Vol. 56, pp. 29–30.

van den Bergh, S., *Galaxy Morphology and Classification*, Cambridge: Cambridge University Press, 1998.

van den Bergh, S., The Early History of Dark Matter, *Publications of the Astronomical Society of the Pacific*, 1999, Vol. 111, pp. 657–660.

Van Helden, A., *Measuring the Universe*, Chicago: University of Chicago Press, 1985.

Villard, R. and Levay, Z., Creating Hubble's Technicolor Universe, *Sky & Telescope*, 2002, September issue, pp. 28–34.

von Humboldt, A., *Cosmos: A Sketch of the Physical Description of the Universe*, Transl. E. C. Otte, Baltimore: Johns Hopkins University Press, 1997.

von Marius, C. F. P., *Der Buch der Palmen*, Cologne: Taschen GmbH, 2016.

Vorontsov-Velyaminov, B. A., *Atlas and Catalog of Interacting Galaxies, Part 1*, Moscow: The Sternberg Astronomical Institute, Moscow University, 1959.

Walsh, D., Carswell, R. F. and Weynmann, R. J., 0957+561 A, B: Twin Quasi-Stellar Objects or Gravitational Lens?, *Nature*, 1979, Vol. 279, pp. 381–384.

Way, M. J. and Hunter, D. (eds.), *Origins of the Expanding Universe: 1912–1932*, Astronomical Society of the Pacific Conference Series, 2013, Vol. 471.

Way, M. J., Dismantling Hubble's Legacy?, in *Origins of the Expanding Universe: 1912–1932*, ed. M. J. Way and D. Hunter, Astronomical Society of the Pacific Conference Series, 2013, Vol. 471, pp. 102–103.

Weinberg, S., Physics: What We Do and Don't Know, *The New York Review of Books*, 2013, Vol. LX, No. 17, p. 86.

Willett, K. W., Lintott, C. J., Bamford, S. P. et al., Galaxy Zoo 2: Detailed Morphological Classification for 304,122 Galaxies From the Sloan Digital Sky Survey, *Monthly Notices of the Royal Astronomical Society*, 2013, Vol. 435, pp. 2835–2860.

Wilson, E. O., *The Diversity of Life*, Cambridge, MA: Belknap Press of Harvard University Press, 1992/Allen Lane, 1993.

Wolf, M., Die Klassifizierung der kleinen Nebelflecken, *Publikationen des Astrophysikalischen Institute Königstuhl-Heidelberg*, 1908, Vol. 3, pp. 109–112.

Wolf, M., On the Dark Nebula NGC 6960 (Über den dunklen Nebel NGC 6960), *Astronomische Nachritchten*, 1923, Vol. 219, pp. 109–116.

Woltjer, H., Spiegelsystems streifenden Einfalls als abbildende Optiken für Röntgenstrahlen (Glancing Incidence Mirror Systems as Imaging Optics for X-rays), *Annalen der Physik*, 1952, Vol. 445, pp. 94–114.

Worthington, A., *The Splash of a Drop*, London: Society for Promoting Christian Knowledge, 1895.

Wray, J., *The Color Atlas of Galaxies*, Cambridge: Cambridge University Press, 1988.

Zwicky, F., Die Rotverschiebung von extragalaktischen Nebeln, *Helvetica Physica Acta*, 1933, Vol. 6, pp. 110–127.

Zwicky, F., On the Masses of Nebulae and of Clusters of Nebulae, *The Astrophysical Journal*, 1937, Vol. 86, pp. 217–246.

Zwicky, F., Luminous and Dark Formations of Intergalactic Matter, *Physics Today*, 1953, Vol. 6, pp. 7–11.

Zwicky, F., Multiple Galaxies, *Handbuch der Physik*, 1959, Vol. 53, pp. 344–385.

Index